AF865203

Find Christ
Find Peace

For the Wounded, the Weary, and the Warrior: From the Trenches of Trauma to the Pages of Peace

This is a rescue mission for the soul

Andrew J. Wein-Wahl

Copyright © 2025 Andrew J. Wein-Wahl

All rights reserved.

No part of this book may be reproduced, distributed, or transmitted in any form or by any means, including photocopying, recording, or other electronic or mechanical methods, without the prior written permission of the publisher, except in the case of brief quotations used in critical reviews, articles, or references.

Scripture quotations are taken from the Holy Bible, New International Version® (NIV®), English Standard Version® (ESV®), and New King James Version® (NKJV®), and are used by permission. All rights reserved by their respective publishers.

ISBN: 978-1-970541-02-1

Dedication

This is for the fallen, those we lost in combat, and those we lost to the war that followed them home. Their names are not forgotten. SSG Adkinson, SGT Alcaraz, SPC George, SPC Page, and far to many that have lost the battle at home; their sacrifice is not in vain. Some gave all on foreign soil. Others lost the battle in their minds, long after the uniforms came off. We honor them with more than memory, we honor them by telling the truth, by healing, and by helping those still in the fight.

To the veteran who is still struggling, still waking up in a fight, still battling the darkness alone, this book is for you. You're not forgotten. You're not weak. You're not beyond saving. There is a way out, and it leads to peace, not the kind the world gives, but the kind only Christ can offer. There is still hope. There is still healing. You are not alone.

To my wife, Rayanne, this journey would not exist without you. From our first connection as kids to the long years apart and the miracle of reunion, your love has been a lifeline. You prayed for me when I didn't have the strength to pray for myself. You believed in me when I didn't believe in anything. Through trauma, addiction, war, and redemption, you've stood beside me, not as someone who had it all together, but as someone who clung to Christ with me. Your story is its own testimony of grace and healing, and your faith helped lead me to mine. This book is as much yours as it is mine.

If you or someone you know is in crisis, please reach out. The Veterans Crisis Line is available 24/7. Dial 988 and press 1, or text 838255 to speak with someone who understands.

Table of Content

Acknowledgments

First and foremost, I want to thank my wife, Rayanne, the love of my life and the truest example of grace and perseverance I have ever known. You have walked beside me through the deepest valleys, believing for light when all I could see was darkness. Your prayers carried me when I had no words left. Your strength, love, and unwavering faith in Christ have been a constant reminder of God's mercy. You are not just my companion, you are living proof that God redeems and restores. Thank you for standing with me, for praying for me before I even returned home from war, and for helping me rebuild a life centered on Christ.

To the pastors and spiritual mentors who spoke truth into my life when I was too broken to listen, thank you. Your obedience to God's call and your patience with wandering souls like mine made room for redemption. To the leaders who taught me that theology is not just about knowledge but about the lived experience of grace, your words helped me piece together my shattered understanding of God. You pointed me back to Jesus again and again, and I am forever grateful.

To the military leaders and mentors who saw something in me, even when I was rough around the edges, thank you for instilling discipline, courage, and purpose in a kid who had known mostly chaos. Your leadership mattered more than you'll ever know. You gave me a mission when I had lost my way, and through that structure, I began to find strength again.

To my brothers and sisters in arms, especially those who served beside me in Afghanistan, you are the heartbeat of this book. Whether you came home or not, your stories are etched into my soul. To those who lost their battles after the war, through overdose, suicide, or silence, you are not forgotten. Your struggle was real, and your memory lives on in every word of this journey. I wrote this book for you. I carried your names through every chapter, every page. You mattered. You still matter.

To the veteran still fighting, know this: you are not alone. Christ has walked every road with you, even the broken ones. He is not afraid of your pain, your past, or your questions. He is the peace you've been searching for. I found Him in every chapter of my life. I pray you find Him too.

To My Editor and Proofreader Robert B.

I want to express my deepest gratitude to my editor and proofreader, whose diligence, insight, and care helped shape this book into something far more powerful than I could have ever done on my own. Thank you for honoring my voice, preserving the raw emotion of my story, and helping it flow with clarity and purpose. You didn't just correct grammar or polish sentences, you entered into the heart of this project.

Your work mattered more than you know.

Introduction

I never set out to write a book. I set out to survive. From my earliest memories in foster care, through years of addiction, and later as a soldier in Afghanistan, I was searching for peace. I carried wounds I didn't have words for, wounds of abandonment, war, and failure. For years I tried to numb the pain with substances, anger, and isolation, but peace was always just out of reach.

What changed my story was not my own strength, it was Christ, and not just Christ as an idea, but Christ alive, present, and revealed in the pages of Scripture. I began to see that every book of the Bible points to Him. From Genesis to Revelation, He was there, whispering hope, shining light in the darkness, and showing me that He is the peace I could never earn or build on my own.

I didn't come to this truth in a straight line. It was through stumbling, failing, and getting back up again. It was through sleepless nights haunted by memories of war, through broken relationships I thought would never heal, and through seasons of shame so heavy I wondered if God could even love someone like me. I remember lying awake in a tent overseas, clutching my rifle tighter than I ever clung to faith, wondering if I'd even make it home. Years later, I remember sitting in silence on the edge of my bed, clouded by smoke and despair, convinced that maybe the world would be better off without me. In those very places when I finally had nothing left to cling to, the Lord met me. He showed me that His grace is greater than my past, His power stronger than my pain, and His peace deeper than my chaos.

This book is for the veteran who came home but still feels lost in the battlefield of the mind. It is for the first responder who carries unseen burdens from the calls nobody else can forget. It is for the foster child who wonders if they'll ever belong, and for the addict who believes they are too far gone. It is for the spouse who feels forgotten, the parent weighed down with guilt, and the one sitting alone wondering if life is still worth living. It is for anyone who feels broken, trapped, or abandoned.

One line from this book says it best:

"Some battles don't make the news. Some wars rage long after you come home."

I here to tell you there is hope. The same Jesus who met me in my lowest moments is the Jesus who walks with you. He is present in every chapter of Scripture, and He longs to bring peace to every chapter of your story.

This is not a book of easy answers. It is a journey through the Bible, through pain, and ultimately into the presence of Christ, where real peace is found. My prayer is that as you read these pages, you will discover what I did: Christ is not only in the Scriptures, He is in your story, and in Him, you can finally find peace.

Andrew J. Wein-Wahl

Chapter 1

Genesis

The Beginning, The Battle, and The Rescue Plan

Chapter 1. Genesis

The Beginning, the Battle, and the Rescue Plan

"In the beginning, God created the heavens and the earth..." (Genesis 1:1)

Before I was out of elementary school, I already knew what chaos felt like.

The kind of chaos that doesn't just happen on the battlefield happens in basements filled with smoke, in foster homes with empty beds, in school hallways where a little boy pretends he's okay but feels utterly unseen. I didn't need someone to explain brokenness to me. I was raised in it.

So, when I read the opening words of Genesis, they don't sound abstract or poetic. They sound like a military commander's voice cutting through confusion: "In the beginning, God created..." Before the noise, before the pain, before the trauma, God, and not just any God, but Elohim the Triune Creator who speaks into nothingness and births light.

This first chapter of Scripture isn't just the beginning of the Bible; it's the beginning of your story, my story, and the story of every soldier, every survivor, and every soul who has ever longed for peace but woke up in war.

Out of Chaos

Genesis 1:2 says, "Now the earth was formless and empty, darkness was over the surface of the deep..." Sound familiar? Formless. Empty. Covered in darkness.

That's what trauma feels like.

That's what addiction does. That's what coming home from war feels like when the battlefield follows you back. When your memories scream louder than your surroundings. When you sit in church but can't feel the presence of God. When you hold your child but can't stop shaking.

Genesis doesn't stop with chaos.

"And the Spirit of God was hovering over the waters."

Even in the darkness, God's Spirit was present. Hovering. Waiting. Preparing to speak, and when He does, everything changes.

The Word Who Was There

We're not left to wonder who this God is. John 1:1 tells us plainly: "In the beginning was the Word, and the Word was with God, and the Word was God." That Word is Christ.

Christ was not absent in the beginning. He was there. He is the beginning. He was the One through whom all things were made (John 1:3), including the very breath you're using to read this now.

That means your life isn't a cosmic accident. Your trauma isn't random. Your suffering isn't unseen.

Jesus was not an afterthought. He wasn't "Plan B." He was the Lamb slain from the foundation of the world (Revelation 13:8). From the first page of Genesis, God had already written a rescue mission, with your name etched into the plan.

The First Promise of Hope

In Genesis 3:15, we hear the first whisper of the Gospel: "I will put enmity between you and the woman, and between your offspring and hers; he will crush your head, and you will strike his heel."

This isn't mythology. This is military strategy. This is battlefield language.

God is declaring war on evil, promising a future where the Seed of the woman, who is Christ will crush the serpent. He'll take a hit in the process, but He'll win. He'll bleed, but He'll rise. He'll die, but He'll overcome.

For every soldier who's lost a battle. For every addict who relapsed. For every parent who buried a child. This verse is not just hope, it's a battle cry.

The Gospel Echoes in Genesis

The beauty of Genesis is that the Gospel isn't limited to one verse. It reverberates through the whole book:

> In Abel, we see a sacrifice accepted because of blood, not because of performance. That's grace.
>
> In Noah, we see salvation through a single door—just as Jesus said, "I am the door" (John 10:9). There's only one safe place in the storm: inside the presence of Christ.
>
> In Abraham, we learn that faith isn't about having it all together, it's about trusting the One who does.
>
> In Isaac, we see the beloved son laid down by his father. Only this time, on Calvary, there was no ram in the thicket. Jesus was the substitute.
>
> In Joseph, we see the one who was betrayed, falsely accused, imprisoned, and forgotten… only to be lifted up to save the very ones who betrayed him. That's not just Joseph's story, that's Jesus'.

For us veterans, first responders, broken people, Joseph's story reminds us that, even when the world forgets you, even when your name is buried under pain, God remembers, and He restores.

Trauma and the Triune God

Genesis doesn't shy away from trauma. It's filled with it.

From Cain's murder of Abel to Abraham's near sacrifice of Isaac… from Sodom's destruction to Jacob's deceit… from Rachel's infertility to Joseph's betrayal, Genesis is raw, but so is life.

God isn't afraid of the mess. He doesn't avoid broken families or dysfunctional stories. He steps into them. He works through them, and He redeems them.

That's why Genesis is a book for those of us who've lived through hell and, somehow, still believe heaven isn't just a fairy tale.

"The Lord is near to the brokenhearted and saves the crushed in spirit." Psalm 34:18

A Soldier's View of Genesis

Genesis resonates with me not because it's ancient but, because it's honest.

When I was deployed in Afghanistan with the 173rd Airborne, we had our own beginnings and battles. We saw light and dark, life and death. We watched friends bleed and enemies fall. We walked in deserts both literal and spiritual.

Genesis reminds me of the battlefield. There's no polished surface here just soil, sweat, and sacrifice.

But in every war zone, there's also hope. There's the sound of a chopper coming in. There's the moment the medics arrive. There's the voice over the radio saying, "You're not alone."

That's Genesis.

God's first words aren't judgment, they're life. His first act isn't punishment, it's creation. His first promise isn't condemnation, it's redemption.

Christ from the Beginning

You don't have to wait until Matthew to find Jesus. He's there when Adam hides in shame. He's there when Sarah laughs in disbelief. He's there when Hagar cries in the wilderness. He's there when Jacob wrestles in the night. He's there when Joseph weeps in prison. He's been there for you, too.

The Long View

God doesn't rush healing. He works through process, people, and pain.

Abraham didn't see the full promise. Joseph waited decades, and yet, God's plan never failed.

Sometimes your life may feel more like Genesis 3 than Genesis 1. But hang on, Genesis 50 is coming. "What you meant for evil, God meant for good…" (Genesis 50:20)

Chapter 2

Exodus

Chapter 2. Exodus

Hope for the Warrior's Soul

"You yourselves have seen what I did to Egypt, and how I carried you on eagles' wings and brought you to Myself." — *Exodus 19:4*

From Slavery to Freedom

The book of *Exodus* is the story of deliverance from slavery to freedom, from Pharaoh's grip to the presence of God. It is the story of a God who sees, hears, and comes down to rescue His people. For many of us who've walked through fire, combat, trauma, addiction, and guilt, Exodus feels like more than history. It feels personal. It feels like our story too.

At the center of Exodus is a confrontation, not just between Moses and Pharaoh, but between Yahweh and every false god, every oppressive power, and every spiritual stronghold that says, "You don't belong to God, you belong to me." For veterans, that voice might come from past mistakes, combat scars, survivor's guilt, or addiction, but the God of Exodus speaks louder. He responds with fire, with plagues, with unrelenting grace, and most importantly, with His name.

The Name that Breaks Chains

In Exodus 3, God reveals Himself to Moses as "I AM WHO I AM." Yahweh. It's not just a title; it's a declaration of absolute presence and power. This is the voice that speaks to broken men in deserts. This is the same voice Jesus echoed when He said, "Before Abraham was, I AM" (*John 8:58*). This is the God who doesn't wait for us to clean up. He enters the pit, the pain, and the PTS (Post Traumatic Stress), and pulls us out.

The Blood That Covers

The defining moment of Exodus is the Passover, when a lamb is sacrificed and its blood is placed on the doorposts. That night, death passed over every house covered by the blood. That was no coincidence. It was a combat strategy from Heaven, a preview of the Cross. Jesus is our Passover Lamb, sacrificed so that the death we earned would pass over us.

He doesn't just get us out of Egypt. He brings us out of sin, shame, and spiritual bondage. He leads us out of the trauma and teaches us to walk in freedom.

The Way Through the Waters

At the Red Sea, with Egypt closing in and no way forward, we see Christ as the One who makes a way where there is no way. Trapped between an army and the sea, God splits the water. Jesus would later say, "I am the Way." He doesn't just part waters. He *is* the path to freedom. He walks veterans out of combat zones, trauma cycles, and suicidal thoughts, and He leads them into life again.

Bread in the Wilderness, Water from the Rock

In the wilderness, we see God's provision everywhere.

Manna from heaven falls daily, and Jesus says, "I am the Bread of Life." Water gushes from the rock, and Paul tells us in *1 Corinthians 10:4* that "the Rock was Christ." The Tabernacle is designed in excruciating detail, and every part of it points to Jesus, our once-for-all sacrifice, our Intercessor, our High Priest, and our access to the very presence of God.

For those who've wandered in spiritual deserts, Exodus reminds us that Jesus is our provision, our protection, and our presence.

From Rescue to Relationship

But Exodus isn't just about getting out of Egypt. It's about learning who God is in the wilderness. It's about being reshaped after trauma. It's about learning to trust again, worship again, and live again. It's about becoming a people, veterans, spouses, civilians, survivors, who walk in the presence of God.

Christ is not just the One who saves us. He is the One who walks with us after the rescue. He doesn't lead us out to leave us alone. He continues to be our Shepherd, our Healer, and our Companion through the recovery journey.

He's still with us, leading us from captivity to freedom, from shame to identity, from isolation to brotherhood, from survival to resurrection. That's Exodus. That's the Gospel. That's hope for the warrior's soul.

From Chains to Calling

Many of us know what it feels like to be trapped by addiction, trauma, guilt, shame. Sometimes the chains are visible. Sometimes they're wrapped around our hearts, but the story of Exodus tells us that God doesn't leave us there. He sees our suffering. He hears our cries. And He comes down.

Here's the part we often forget about deliverance is not the end of the story. Once the Israelites were free, they had to learn how to live as the people of God. Freedom brings responsibility. Redemption brings calling, and you, too, were not just saved from something. You were saved for something.

Christ didn't just deliver you to wander. He delivered you to worship. He still walks beside you on the road ahead.

Chapter 3

Leviticus

Christ, Our Holy High Priest and Cleansing Redeemer

Chapter 3. Leviticus

Christ, Our Holy High Priest and Cleansing Redeemer

"For the life of a creature is in the blood… it is the blood that makes atonement for one's life." — *Leviticus 17:11*

The Book We Tend to Skip

Leviticus is the book most people skip. It's filled with sacrifices, blood rituals, rules about skin diseases, and priestly duties. At first glance, it reads more like an ancient field manual than a life-giving message, but when we slow down, when we read it not as a list of regulations but as a whisper of something greater, Leviticus becomes a holy echo, calling out the holiness of God and pointing us to the sacrifice and priesthood of Christ.

To the Israelites in the wilderness, Leviticus wasn't just law, it was life. It was survival in the presence of a holy God. The Tabernacle sat in the middle of the camp, a visible reminder that God dwelled among them, but His holiness wasn't to be taken lightly. Every instruction in Leviticus taught Israel how to live in proximity to a consuming fire without being burned, and at the center of it all was a truth that runs through every chapter: you can't enter God's presence without blood.

That sounds harsh especially to modern ears, but it's also hopeful. Because that blood doesn't have to be ours. From the burnt offerings to the sin offerings, the peace offerings to the guilt offerings, Leviticus shows us that God made a way. and every single time, that way points to Jesus.

Every Offering a Foreshadowing

The burnt offering speaks of complete surrender. It was consumed entirely on the altar, nothing held back. That's Christ, laying down everything, body, soul, and blood for us.

The sin offering carries the weight of guilt. It's not just about mistakes; it's about atonement, about someone else taking the hit so we can go free. That's Christ, stepping into the line of fire.

The peace offering is communion, a shared meal of reconciliation between God and man. It's an invitation to fellowship, and Jesus is the table, the wine, the bread, and the host.

Then there's the scapegoat, sent into the wilderness bearing the sins of the people. That goat walked away with their shame, and Christ walked to the Cross with ours.

The Priest Who Walks With Us

It wasn't just the sacrifices. It was the priests. The high priest wore the weight of the people, literally. The breastplate he carried had twelve stones, one for each tribe. He went before God as their representative. Once a year, he entered the Holy of Holies with trembling hands and the blood of an animal, hoping God would accept the offering.

Jesus is so much more than that priest. He is the Great High Priest who entered once, not with the blood of bulls or goats, but with His own blood, shed on a Roman cross. He didn't just represent the people, He became one of us, and He didn't just make an offering, He was the offering.

The veil that separated man from God was torn the moment Jesus died. What Leviticus made temporarily possible through ritual, Jesus made eternally possible through grace.

When You Don't Feel Clean

Growing up, I didn't feel clean. I didn't feel holy. I didn't feel wanted. I carried the stains of my own sin and the sins of others. Abuse, foster care, drugs, violence, war, guilt, it stuck to me like body armor I couldn't take off.

There were nights I wondered if God would even want someone like me. There were moments I didn't think I'd ever be clean. But then I met Jesus, not just the gentle Savior in paintings, but the High Priest who wades through blood and fire to stand with the broken. and I realized that Leviticus isn't a story of judgment, it's a story of a God who makes a way.

I didn't have to get clean to come to Him. I came to Him, and He made me clean.

Holiness in the Wilderness

Holiness isn't for people who have it all together. It's for people being put back together. The wilderness isn't the absence of God, it's often where we meet Him most intimately.

I've seen the inside of a crack house. I've sat in jail. I've walked through the dust of Afghanistan, my boots pounding the ground while my soul carried weight I didn't know how to name, and yet, in every place I thought was too filthy for God, He was already there.

In the military, we lived by codes and standards. Some of those shaped us for the better. Others left scars. After combat, after addiction, after all the ways we learn to survive, we don't just need discipline, we need healing. That's what holiness is: not cold purity, but intimate healing.

God doesn't say, "Get it together." He says, "Come close. Let Me do the cleansing." Like the priests in Leviticus, we don't purify ourselves, we receive purification.

Learning a New Way to Live

When I came home from the war, I had to unlearn some things. You learn to scan rooftops. You sleep with one ear open. You don't trust easy. You numb the pain. You isolate. You protect.

Leviticus shows us a God who says, “Let Me re-train you. Let Me show you what it means to live in My presence.” That takes time. It takes trust, but it begins with the blood of a sacrifice already made.

Holiness isn’t about cleaning up so we can belong. It’s about belonging to the One who cleans us up. That’s grace. That’s Jesus. and that’s the hope of every veteran, every addict, every spouse who feels distant, every soul who wonders if they’ve gone too far. You haven’t. The Priest is waiting.

Holiness Isn’t a Reward—It’s a Relationship

You don’t earn your place with God, you receive it. Holiness is not behavior first, it’s presence first. The more we walk with Jesus, the more His life becomes ours. We don’t fight our way to purity. We surrender our wounds, our habits, our fears and we walk into the presence of the God who heals.

Let God redefine holiness for you. Let Him teach you that holiness doesn’t mean pretending to be okay, it means walking with the One who is. He is our Great High Priest. He is our final sacrifice, and He has already declared you clean.

Chapter 4

Numbers

Christ, Our Guiding Light Through the Wilderness

Chapter 4. Numbers

Christ, Our Guiding Light Through the Wilderness

Between Rescue and Promise

The book of *Numbers* is about life in the middle. The high of deliverance from Egypt has faded, and the joy of the Promised Land is still out of reach. For the Israelites, and for many of us, the wilderness is the space in between the waiting, the wandering, the wondering. That's where the real battle begins.

This book is about marching orders, rebellion, discipline, and divine patience. It's not flashy, but it's raw. It's honest, and it speaks directly to the veteran, the addict, the trauma survivor, the wounded, and the weary. Anyone who's not where they used to be, but not yet where they hope to be. That's where I found Christ.

The Wilderness Reveals the Heart

For Israel, the desert wasn't just sand and silence. It exposed everything hidden beneath the surface fears, idols, and deeply rooted beliefs shaped by slavery. They had been delivered, but they still thought like captives. The wilderness revealed their true condition. That's what trauma does too. After the war, after the addiction, after the crisis, we enter a wilderness of the soul. The structure is gone. The mission is unclear. Our sense of self feels fractured. I didn't meet Jesus on a mountaintop, I met Him in that dry, empty place where my past haunted me, and my future felt like a question mark. That was my Numbers.

Christ in the Wilderness

Jesus is all over this book. He is the Cloud by day and the Fire by night, leading with visibility and consistency. He is the Rock that poured out water, the Bread that fell from heaven, and the Serpent lifted up for healing.

He didn't leave Israel in their confusion, and He doesn't leave us in ours. Even when they rebelled, He was still present. Even when they failed, He remained faithful. The wilderness didn't destroy them, it refined them. It was the place where God transformed survivors into sons and daughters.

Boot Camp for the Soul

The military gave us discipline, structure, and survival strategies. We followed SOPs (Standard Operating Procedures), respected rank, and lived by the uniform code of military justice or UCMJ, but not everything we learned served us once we came home. Some habits saved our lives. Others numbed our souls. The uniform came off, but the tension remained.

The wilderness is where God begins to unlearn us from survival and retrain us for relationship. It's where we learn to depend on grace, not performance. It's where we lay down our weapons and pick up our identity. We don't need to fight for approval anymore we're already claimed.

Mighty Oaks and the Power of Identity

At Mighty Oaks Warrior Programs, I learned something that changed my life: identity is not earned, it's received. In Christ, we are not defined by trauma, mistakes, or our military records. We are sons and daughters of the living God, but like Israel in the wilderness, we often have to walk through some fire before we believe it. The battle shifts. It's no longer about taking ground; it's about trusting God. Not about proving strength, but receiving grace. The wilderness becomes a proving ground, not of our performance, but of God's faithfulness.

Maybe you're in that place right now, not who you used to be, but unsure of who you're becoming. You feel tired, disoriented, even forgotten. Hear this: Christ has not left you. He is still leading you, still feeding you, still forming you. The middle isn't meaningless. It's the place of deepest transformation.

Breaking Ranks with Old Chains

Many of us were trained to shut down, avoid weakness, and keep marching no matter the cost. Those rules helped us survive war, but they often sabotage our peace. Like Israel, we've been delivered from bondage, but bondage still echoes in our hearts.

Christ doesn't just clean us up, He makes us new. He gives us a new name, a new mission, and a new way to live. You're not in survival mode anymore. You're in transformation, and you're not walking alone. He is with you, every step of the way.

Chapter 5
The Wilderness Within

Trauma, Suicidal Thoughts, and the Fight for Life

Chapter 5. The Wilderness Within

Trauma, Suicidal Thoughts, And the Fight for Life

"The Lord is close to the brokenhearted and saves those who are crushed in spirit." — Psalm 34:18

Some battles don't make the news. Some wars rage long after you come home.

For many of us who served, whether in Afghanistan, Iraq, or elsewhere, the battlefield followed us back. We carried it in our memories, our muscle tension, our midnight thoughts. We walked through airports like strangers. We sat in church feeling unworthy. We lay in beds wondering if anyone even noticed we made it home.

According to the Department of Veterans Affairs, an estimated 17–22 veterans take their own lives each day in the United States. Some say the number is even higher when you count unreported or misclassified cases. That's more than just a statistic, it's a silent war. A generation of warriors quietly falling in the wilderness. And the wilderness is real.

I Know That Wilderness

When I came home from Operation Enduring Freedom X (OEF X), with the 173rd Airborne, the external fight may have ended, but an internal one began. I didn't know how to explain the noise in my head, the flashbacks in my sleep, or the guilt I carried in my soul. I couldn't talk about it in church because I wasn't going, and I didn't know if God was even listening anymore. I saw too much, had done too much and lost too much.

They train you how to fight. They don't always train you how to come home.

The trauma took many forms moral injury, survivor's guilt, anger, addiction, numbness. Then there was another wound, one I never saw coming. While I was in Afghanistan, my 2nd marriage fell apart. My then wife cheated on me while I was downrange. In a place where I already questioned my worth and purpose, that betrayal cut deep. It wasn't just the end of a relationship, it was the shattering of trust, the loss of a home I thought I was fighting to protect. I was carrying the war on my shoulders and came home to find the battlefield had followed me into my own living room.

I smiled when I needed to, but I was dying inside. I numbed the pain any way I could. I started to believe the lie that maybe the world would be better off without me.

If that's you, I see you. More importantly, Jesus sees you.

Jesus in the Dark

One of the most powerful moments in Scripture comes from a man we don't usually talk about in this context: Elijah. After a great victory, Elijah ran into the wilderness and begged God to end his life.

"I have had enough, Lord," he said. "Take my life." (1 Kings 19:4)

It wasn't weakness. It wasn't unbelief. It was exhaustion. But what did God do? He didn't rebuke Elijah. He sent an angel to feed him. He gave him rest. He reminded him of his calling, and

He revealed Himself not in the earthquake, not in the fire, but in a still small voice.

That's how Jesus meets veterans in the wilderness.

Not always in thunder, but in the quiet whisper:

"You are still Mine. You still matter. I'm not done with you."

The Gospel Is for the Wounded

One of the lies trauma tells us is that we're too far gone. That no one could possibly understand what we've been through, but here's the truth:

Jesus was betrayed by His closest friends. He was beaten, mocked, stripped, and humiliated. He carried the weight of sin and shame, not just physically, but spiritually. He was abandoned, and He cried out, "My God, My God, why have You forsaken Me?"

Jesus understands trauma, not from a distance, but from within it. The Cross was not clean. It was raw, brutal, and personal. He bore it so you wouldn't have to bear yours alone.

Through His death, He took on our darkness. Through His resurrection, He gave us permission to live again.

From Trigger to Testimony

Healing doesn't happen all at once. It's not always linear either. Some days I still hear echoes of war in my spirit. What used to trigger fear, anger, frustration, and despair now reminds me of grace.

At "Mighty Oaks Warrior Programs", I learned something lifechanging: My identity is not in my past, not in my pain, not even in my performance. My identity is in Christ.

I'm not a victim. I'm not a statistic. I'm a son of the Living God, and so are you.

You may carry scars. But those scars can become signs of hope for others still bleeding.

So What Do You Do When You Feel Like Giving Up?

1. Speak the Name of Jesus

There is power in His name. In the middle of the night, in the middle of a flashback, in the middle of a breakdown, whisper His name, He's already nearby.

2. Tell Someone the Truth

Isolation is a liar, don't believe it. Tell someone, a chaplain, a friend, a counselor, a brother in Christ. Let the light in.

3. Stop Trying to Be Strong Alone

Even Jesus had help carrying His Cross. You don't have to carry yours by yourself. Strength is admitting you're not okay and accepting help anyway.

4. Find Your Tribe

Programs like Mighty Oaks, Celebrate Recovery, the VFW, and American Legion to name a few exist for a reason. You need people who've been through the fire and lived to tell the story. You weren't made to fight alone.

5. Get in the Word

Not as a checkbox, but as a lifeline. Scripture isn't just information, it's oxygen. Let the words of life push back the lies of death.

A Letter from One Warrior to Another

"I will not die but live, and will proclaim what the Lord has done."— Psalm 118:17

To the brother or sister staring into the darkness. To the one with too many memories and not enough hope. To the one who's been through hell and wonders if you'll ever feel whole again

I know where you are. I've sat in that silence. I've felt that weight. I've heard that whisper: "Maybe they'd be better off without me."

Let me speak truth over that lie.

"The thief comes only to steal and kill and destroy; I have come that they may have life, and have it to the full." — John 10:10

The enemy of your soul wants you to believe that your pain is final. That your trauma defines you. That your best days are behind you. Jesus came to give you life, not just survival, not just numbness, but abundant, meaningful life.

I served in Afghanistan with the 173rd Airborne. I've seen things that left marks I couldn't explain. I carried those things back home, and they nearly took me out. PTS. Guilt. Addiction. Despair. I even questioned God's presence, His goodness, His love.

"Even though I walk through the valley of the shadow of death, I will fear no evil, for You are with me." — Psalm 23:4

He was there. Not always in the ways I expected. Not always with answers, but always with presence. Always with mercy. Always with scars of His own to show me I wasn't alone.

I found healing in Christ. Not a magic fix. But a slow, steady resurrection. One breath at a time. One Scripture at a time. One prayer at a time.

"He heals the brokenhearted and binds up their wounds."

— Psalm 147:3.

"The Lord is close to the brokenhearted and saves those who are crushed in spirit." — Psalm 34:18

You are not a lost cause. You are not your discharge paper, your mistake, your addiction, your diagnosis. You are not your past.

In Christ, you are Forgiven (1 John 1:9), Chosen (John 15:16), Adopted (Ephesians 1:5), More than a conqueror (Romans 8:37), A new creation (2 Corinthians 5:17).

The Cross wasn't just about saving you from hell. It was about saving you from the hell inside you right now. Jesus already bore the shame. You don't have to carry it anymore.

"Come to Me, all you who are weary and burdened, and I will give you rest." — Matthew 11:28

Take Him up on that. Call a brother. Call Mighty Oaks. Call on Jesus. Just don't end your story. Not like this.

"For I know the plans I have for you," declares the Lord, "plans to prosper you and not to harm you, plans to give you a future and a hope." — Jeremiah 29:11

There's more. There's still a fight worth fighting, but now, it's for your soul, and you don't fight alone.

I'm still here because of Jesus, and because someone reached out when I didn't think I was worth saving. This is me reaching out to you. Hold on. Let God hold you. There's a new chapter waiting to be written.

Andrew J. Wein-Wahl

You're Not Alone — Help Is Here

"Carry each other's burdens, and in this way, you will fulfill the law of Christ." — Galatians 6:2

You don't have to suffer in silence. The war you're fighting now can be won—but not alone. There are people who care. People who understand. People who've walked through the fire and come out alive—and want to walk with you.

Talk to Someone Right Now

•Veterans Crisis Line (24/7 Confidential Support):

Dial988, then press 1

•Or text 838255

•www.veteranscrisisline.net

You can also message me at kingdomstories.org@gmail.com for more information. You can also check out my website at www.kingdomstories.org for a full list of places you can find help.

Chapter 6

Deuteronomy

Christ, The Fulfillment of the Law and the Voice of Fire

Chapter 6. Deuteronomy

Christ, The Fulfillment of the Law and the Voice of Fire

"The LORD your God will raise up for you a prophet like me from among you, from your fellow Israelites. You must listen to him." — Deuteronomy 18:15

Deuteronomy is a book of remembrance. A last word before a new beginning. The Israelites stand at the edge of the Promised Land, but Moses knows he won't enter with them. What he gives them instead is something deeper: a vision of faithfulness, a renewed covenant, and a call to choose life.

It's also a book of heart level obedience—not just external conformity.

"Love the LORD your God with all your heart and with all your soul and with all your strength." — Deuteronomy 6:5

This is the heartbeat of the entire book. And it's the same verse Jesus quoted when He was asked what the greatest commandment was (Matthew 22:37). In this way, Deuteronomy brings the law and the Gospel together, it doesn't just tell us what to do, it shows us who to become, and it points forward to the One who would do it perfectly: Jesus Christ.

Jesus in Deuteronomy

Moses tells the people that God will raise up a prophet like him. Centuries later, Peter (Acts 3:22) and Stephen (Acts 7:37) both confirm that prophet is Jesus.

Just as Moses interceded for the people, Jesus stands between us and judgment. Just as Moses went up the mountain to hear from God, Jesus is the Word of God made flesh. Just as Moses gave the law, Jesus fulfills the law and writes it on our hearts.

"Do not think that I have come to abolish the Law or the Prophets; I have not come to abolish them but to fulfill them."

— Matthew 5:17

Christ is the true and better Moses, not only leading us out of slavery but empowering us to walk in the Spirit, not just by rules, but in relationship.

Choose Life

"This day I call the heavens and the earth as witnesses against you that I have set before you life and death, blessings and curses. Now choose life…" Deuteronomy 30:19

As a veteran, this hits hard. I've stood in the presence of death. I've watched lives end in a moment. I've faced the decision to live or die—not just physically, but spiritually and emotionally.

There came a moment when I had to stop surviving and start living. Not numbing, not escaping, not performing, but living free, forgiven, and full of the Spirit of God.

That choice is still on the table for every one of us. Jesus is Life (John 14:6). To choose Him is to choose wholeness, identity, purpose, healing, peace.

Don't Miss the Mountain

One of the most sobering moments in Deuteronomy is when Moses climbs Mount Nebo. From the summit, he sees the land God promised. He can see the future, but he won't live in it. He dies on the mountain, faithful but unfulfilled.

There was a time in my life when I could see the possibility of change, but I didn't know how to get there. I could picture the Promised Land of peace, freedom, family, healing, but it felt out of reach. My trauma told me I didn't belong there. My past whispered that I wasn't worthy to enter. I stood on the ridge line, looking in, feeling like my life had been reduced to watching other people walk in while I stayed on the outside.

Maybe you've been there standing at the edge of what God promised, feeling the weight of your own history pressing you back. Maybe you've even convinced yourself you'll never cross over. That's what the wilderness can do, it convinces you that what's ahead is only for others, not for you.

Jesus changes everything. He doesn't just show you the land, He leads you in. He doesn't just point out the promise He fulfills it. Here's the truth that finally broke through my own doubts: in Him, I've already made it. The crossing isn't in question. The victory isn't a maybe. The Promised Land isn't just a dream for the "better" Christians, it's an inheritance that's already mine because I belong to Him.

Chapter 7
Joshua

Christ Our Commander and Conqueror

Chapter 7. Joshua

Christ Our Commander and Conqueror

"Have I not commanded you? Be strong and courageous. Do not be afraid; do not be discouraged, for the LORD your God will be with you wherever you go." — Joshua 1:9

When the book of Joshua begins, Moses is dead. The wilderness season is over. Now it's time to move forward, to fight, to take hold of the promise, but they don't do it with their own strength. They don't charge into battle without orders. They follow a Commander who goes before them, Yahweh Himself.

Joshua 1:9 is one of my favorite verses in all of Scripture. In fact, it was one of the first I ever memorized. I held onto it like a lifeline through dark nights and through battles in my own mind. "Be strong and courageous" isn't just a motivational pep talk. It's a command backed by the promise that the Lord your God is with you wherever you go. That changes everything.

I've had times when I thought being strong meant gritting my teeth and muscling through on my own. This verse reminds me that my strength isn't something I generate, it's something I receive. I can be strong because He is strong. I can be courageous because He goes before me, stands beside me, and guards my six. The pressure isn't on me to win the war, it's on me to trust the One who already has.

That's where we meet Christ in Joshua, not as the suffering servant, not as the sacrificed lamb, but as the Captain of the LORD's army (Joshua 5:13–15). When Joshua encounters a divine warrior standing with sword drawn, he asks, "Are you for us or for our enemies?" The answer is unexpected: "Neither, but as Commander of the army of the LORD I have now come." It's not about getting God to choose our side, it's about choosing His.

As veterans, we know the chain of command is only as effective as the trust we have in the one leading us. In Christ, we have a Commander who never loses a battle, never abandons His troops, and never sends us anywhere He hasn't gone first. He's not only our strategist, He's our shield.

There are still battles to fight, strongholds to tear down, Jericho's to march around, and rivers to cross, but the outcome is not uncertain. The same God who told Joshua to be strong and courageous is telling us the same thing today. He's giving us the same reason: "For the LORD your God will be with you wherever you go."

Because He is strong, you can be strong. Because He is victorious, you can walk in victory. The Commander of Heaven's army is not just for you, He's with you, every step, every fight, every breath, until the final battle is won.

Before the first battle is fought, there's a river to cross. The Jordan stands between wilderness and promise. It's not parted by Joshua's might, but by obedience. The moment the priests step into the water carrying the Ark, the river stops.

"And as soon as the priests who carried the ark reached the Jordan and their feet touched the water's edge, the water from upstream stopped flowing."— Joshua 3:15–16

The Ark represented God's presence. Today, we have Christ (Emmanuel), God with us, going before us into every battle, every unknown, and every Jordan.

"When you pass through the waters, I will be with you."

— Isaiah 43:2

We are fighting from victory, not for it. The battles that follow Jericho, Ai, and Gibeon teach us something vital: obedience is greater than strategy. God doesn't ask Israel to be perfect warriors; He asks them to trust Him, to listen, and to act in faith. Even when the battle plan doesn't make sense (like walking around Jericho in silence for seven days), or taking rice and crank radios to a village that just killed 4 of your brothers with an IED attack the day before. The victory comes not by might, but by the Spirit.

"Not by might nor by power, but by My Spirit,' says the LORD Almighty."— Zechariah 4:6

As a veteran, this flips the script. In the military, we're trained to dominate, to outthink, to overpower, but in the kingdom of God, the strongest warrior is the one who surrenders to the Lord's command.

I used to think spiritual warfare meant fighting harder. Now I know it means resting deeper in the authority of Jesus. We don't fight for victory, we fight from it. Read that again: we don't fight for victory, we fight from it. How refreshing is that? No matter what we have done or what we are going to do, we stand in a place of victory because of what Christ has done for us.

"The LORD gave them rest on every side… Not one of all the LORD's good promises to Israel failed; everyone was fulfilled."

— Joshua 21:44–45

After the wars, the land is divided and assigned. This part of Joshua can feel dry, but don't miss the point. Every tribe receives an inheritance. And the only one who doesn't? The Levites.

"The LORD Himself is their inheritance."— Joshua 13:33

That's our ultimate inheritance too, not just peace, or power, or heaven, but Christ Himself.

Courage Is a Command

When I came home from Afghanistan, courage looked different. It wasn't running toward gunfire it was showing up to counseling. It was walking into church again. It was picking up the phone when I didn't want to live anymore.

Courage wasn't about being unafraid it was obeying God even while I was afraid. Joshua heard it over and over:

"Be strong and courageous."— Joshua 1:6, 7, 9, 18

Why? Not because he was tough, but because God was with him, and He is with you too.

Chapter 8

Christ in the story of Rahab

Mercy for the Marginalized

Chapter 8. Christ in the story of Rahab

Mercy for the Marginalized

"By faith the prostitute Rahab, because she welcomed the spies, was not killed with those who were disobedient."

— Hebrews 11:31

Before Jericho's walls fell, God was already working inside them. Not in the palace. Not among the leaders. In a brothel.

Her name was Rahab. Her title? Prostitute. Her reputation? Stained. Her house? On the edge of the wall. Her allegiance? About to shift heavenward.

When the spies enter Jericho, they don't find a warrior, they find a woman with a past, and yet, Rahab isn't confused. She tells them plainly:

"I know that the LORD has given you this land… the LORD your God is God in heaven above and on the earth below."

— Joshua 2:9–11

She had heard the stories. She believed the reports. Faith was born in a place most people would overlook.

This woman abused, used, and discarded becomes the vessel through which God protects His people. Because of her faith and obedience, her entire household is saved.

From Scarlet Shame to Scarlet Grace

The spies give Rahab one instruction:

"Tie this scarlet cord in the window through which you let us down…" — Joshua 2:18

The scarlet cord hanging in her window is a sign of mercy. It echoes the blood on the doorposts at Passover, and it foreshadows the blood of Christ that marks every sinner who calls out for mercy. What the world saw as a mark of shame, God transformed into a banner of salvation.

"Though your sins are like scarlet, they shall be as white as snow." — Isaiah 1:18

Rahab isn't just remembered in Joshua, she's honored in the Gospels. Matthew 1 lists her in the genealogy of Jesus. That's right a prostitute from Jericho is an ancestor of Christ. She's also one of only two women mentioned in Hebrews 11, the Hall of Faith.

Let that sink in. Jesus didn't just save people like Rahab. He came from people like Rahab.

Grace for the Disqualified

I've met a lot of warriors who think they're too far gone, too dirty, too broken too addicted and too ashamed. I used to think that about myself.

Rahab's story flips the script. God didn't overlook her He chose her. Not because of her résumé, but because of her faith.

Maybe you feel like you've been living on the edge of the wall, unnoticed, unwanted, and unsure if God could ever use someone like you. Maybe your past screams louder than your hope, but listen, Jesus rewrites stories. Grace chooses the outsider. Mercy calls your name. Rahab's window wasn't her end, it was her beginning.

Prayer: For the One Who Feels Too Far Gone

Jesus,

You see what no one else sees. You step into broken places and call out treasure. You take the overlooked, the ashamed, the outcast, and You write them into the very lineage of salvation.

We praise You for mercy that reaches into brothels, into jail cells, into battle zones, into trauma. No place is too dark for Your light.

Remind us that You don't love us because of what we've done, but because of who You are. Remind us that our past does not cancel our purpose. Remind us that grace is for people like Rahab, and people like us.

Hang the scarlet cord over our stories. Cover us with Your blood. Call us by a new name.

In the name of the Redeemer who came through the house of a prostitute to rescue the whole world, Amen.

Chapter 9

Judges

Christ Our Deliverer in the Days of Darkness

Chapter 9. Judges

Christ Our Deliverer in the Days of Darkness

"In those days there was no king in Israel; everyone did what was right in his own eyes."— Judges 21:25

Judges is the rawest, grittiest book of early Israel's story. It reads like a broken record:

The people forget God. They fall into chaos. They cry out in desperation. God sends a deliverer. Peace returns. They Repeat.

This isn't the book of polished heroes, it's the book of messy, unlikely saviors. The deliverers, judges like Ehud, Deborah, Gideon, and Samson are often deeply flawed, but the point isn't their perfection. The point is that God keeps showing up.

Each judge is a shadow, a temporary rescuer for a specific crisis. But their stories point us forward to the ultimate Judge, the One who doesn't just deliver us from enemies but from ourselves:

***Jesus Christ*.**

Like Othniel, Jesus is filled with the Spirit and brings true peace. Like Deborah, He brings wisdom and prophetic truth. Like Gideon, He calls the weak and says, "Go in the strength you have." Like Samson, He stretches out His arms and brings victory through sacrifice except Christ dies not in failure, but in triumph.

"When the people cried out to the LORD, He raised up for them a deliverer." — Judges 3:9

That's what Jesus does. Over and over again. "Everyone did what was right in their own eyes." Judges ends on this haunting line because there was no king. But spiritually speaking, they had rejected the true King, Yahweh, long before they begged for a human one.

The result? Violence. Abuse. War, and collapse. Judges 19 is one of the most disturbing chapters in the entire Bible in my opinion. It's a brutal reminder of what happens when sin runs unchecked and God's justice is ignored, and yet, into this darkness, God keeps sending light.

That's Christ in Judges. Not center stage, but behind the scenes, working through broken men and women, answering cries, raising up help, and never abandoning His people, even when they abandon Him.

God Uses the Broken

One of the hardest things to believe as a veteran or as anyone carrying shame is that God can use broken people and Judges proves it.

Gideon was anxious and afraid. Barak wouldn't go to battle without Deborah. Jephthah made reckless vows. Samson was ruled by lust and anger and yet, God still worked through them.

"My grace is sufficient for you, for My power is made perfect in weakness." — 2 Corinthians 12:9

You don't have to be cleaned up, qualified, or perfect. You just have to cry out.

That was true for me, too. I didn't grow up in stability. I wasn't raised in a loving church family with consistent support and Sunday School. I was ripped from my mother's arms and shuffled through 14 foster homes before I even had the chance to figure out who I was. Each placement felt like a silent judgment, another reminder that I didn't belong, that I was a problem too heavy to carry, that I was a piece of trash, easily discarded.

In that chaos, I felt far from God. Honestly, I didn't know if He was even real and if He was I was so angry at Him. Looking back now, I see it. He was raising me up. To write this book. He gave me my last foster mother who ended up adopting me. A group of men in the Boy Scouts that mentored me and taught me how to do things men need to know how to do. Even when I didn't call out to Him, He was coming for me.

That's the thing our mess, as veterans or civilians, doesn't scare Him off. Post traumatic stress disorder or PTSD, failed marriages, addictions, moral injuries, nights we'd rather forget, none of it disqualifies us from God's rescue. The same God who delivered Israel from their enemies in the middle of their rebellion will meet us in our own battlefields, even the ones in our minds.

God doesn't wait until the fight is over to step in, He steps into the fight with us. Our brokenness is not a barrier to His presence. If anything, it's the backdrop where His grace shines brightest.

That's the God of Judges, not afraid of the mess, not offended by our wandering, and not shocked by our sin. A God who shows up, again and again and again. So, cry out. Deliverance is never far from the brokenhearted.

Chapter 10

Ruth

Christ, Our Kinsman Redeemer

Chapter 10. Ruth

Christ, Our Kinsman Redeemer

"But Ruth replied, 'Don't urge me to leave you or to turn back from you. Where you go I will go, and where you stay I will stay. Your people will be my people and your God my God.'"

— Ruth 1:16

The book of Ruth opens in a famine and ends in a family. It begins with loss and ends with legacy.

It's only four chapters long, but it holds a whole world of heartbreak, healing, and holy redemption. Most of all, it gives us one of the clearest pictures of Christ as our Kinsman Redeemer, the One who steps in, pays the price, and restores what was lost.

From Moab to Bethlehem: The Journey of the Outcast

Ruth was an outsider. A widow. A foreigner. By every cultural standard, she didn't belong in Israel. And yet, she becomes part of the royal line of David, and of Jesus.

The Gospel isn't about deserving. It's about belonging because of grace.

"Remember that at that time you were separate from Christ… Without hope and without God in the world. But now in Christ Jesus you who once were far away have been brought near by the blood of Christ." — Ephesians 2:12–13

That's Ruth's story. That's our story.

Boaz and the Redeemer's Role

Boaz doesn't just fall in love with Ruth. He redeems her.

According to Israelite law, a kinsman redeemer was a male relative with the right to buy back land, restore family names, and care for widows.

Boaz stands in the gap. He counts the cost. He steps in when no one else will.

He's a foreshadowing of Jesus, our true Kinsman Redeemer.

"The Son of Man did not come to be served, but to serve, and to give His life as a ransom for many." — Matthew 20:28

Where we were outsiders, He made us family. Where we were empty, He made us full. Where we were destitute, He brought restoration.

A Faithful Woman in a Hard World

Ruth isn't just remembered because she ended up in Christ's family tree. She's honored because she walked through pain with loyalty, faith, and sacrificial love.

She stuck by Naomi when she didn't have to. She worked the fields when no one asked her to. She laid herself at Boaz's feet, not in desperation, but in bold humility.

"She is worth far more than rubies." — Proverbs 31:10

Faithfulness in the Fields

The story of Ruth speaks deeply to those who know loss, especially to the wives and partners of wounded warriors.

While the next chapter will focus directly on military spouses, Ruth's story is already speaking to them. It speaks to anyone who has stood by someone shattered by trauma, addiction, grief, or war. Ruth loved a broken family back to life. Not with magic. Not with sermons. But with presence.

I think of those who married a soldier and ended up married to a stranger, someone changed by war, broken by life, distant, angry, and unreachable. I think of those who stayed, even when they had every reason to leave. I think of the Ruths.

If that's you, your story matters. Your loyalty matters. And Jesus sees it.

"God is not unjust; He will not forget your work and the love you have shown Him as you have helped His people and continue to help them." — Hebrews 6:10

Boaz restored Ruth's life. Jesus restores ours. He is the Redeemer of every widow, every spouse, every soul that feels cast aside.

Chapter 11

The Ones Who Waited

Chapter 11. The Ones Who Waited

Christ's Mercy for the Military Spouse

"Love bears all things, believes all things, hopes all things, endures all things. Love never fails." — 1 Corinthians 13:7–8

Not everyone wears the uniform. Not everyone carries a weapon. But some fight a war no one ever sees. They wait. They weep. They pray. They stay.

This chapter is for the military spouse, the one who watches, prays, hopes, and sometimes breaks under the weight of loving someone who came home… but not whole.

When War Comes Home

There's no field manual for loving a broken soldier. There's no chaplain briefing for spouses on how to navigate trauma, addiction, withdrawal, rage, shame, or silence. Some days we come home smiling. Other days, we're ghosts of the men and women you kissed goodbye.

I know this personally. When I was deployed to Afghanistan, my marriage at the time fell apart. My wife then cheated on me while I was downrange. That betrayal cut deep, not just because of what happened, but because it happened when I was already carrying the weight of war. I came home to find that the battlefield had followed me back, right into my own living room.

But that's not the end of my story.

Years later, God brought Rayanne my 9th grade high school girlfriend back into my life. Someone who knew my brokenness, saw my wounds, and still chose to love me. When we got married, the war inside me wasn't over. My healing was and still is a work in progress. Knowing all of that she stood beside me anyway. She prayed for me. She reminded me I was more than my past. She has walked through the long nights, the bad days, and the slow steps forward.

Rayanne has been proof in my life that Christ's mercy is not theoretical, it's personal. She has loved me like Jesus loves His church: fully aware of my flaws and fully committed to staying.

Unseen Wounds, Unshakable Love

The spouse's role is often overlooked. You're not the one getting medals, but you bear the cost. You hold the family together. You pray when no one else sees. You listen to the stories we can't finish. You endure the silence. You absorb the fallout. You love us even when we're unlovable. I believe God is speaking to every spouse who has ever felt invisible. He is telling you I see you. I know. I'm with you.

"The eyes of the LORD are on the righteous, and His ears are attentive to their cry." — Psalm 34:15

Faith Like Ruth, Strength Like Christ

Your story might not make headlines, but it echoes the faithfulness of Ruth, who clung to her mother-in-law and to the God of Israel even after all hope seemed lost:

"Where you go, I will go… Your God will be my God." — Ruth 1:16

That's not just poetry. That's covenant love. If you've stood by a hurting warrior, you know something about covenant. You've stood where others walked away. You've kept believing when others gave up. You've prayed when nothing seemed to change.

Not Forgotten

Maybe you've wondered, "What about me?" Maybe your soul is tired. Maybe you feel like the collateral damage of someone else's war. You are not forgotten. You are not background noise in someone else's calling. You are not invisible to God.

"God is not unjust; He will not forget your work and the love you have shown Him as you helped His people and continue to help them." — Hebrews 6:10

Christ doesn't just redeem soldiers, He redeems families. He binds up not only the wounds of the warrior, but of the one who waited, wept, and hoped through every storm.

"Blessed are those who mourn, for they shall be comforted." — Matthew 5:4

To every spouse who stayed even when it was hard, know this, your endurance reflects the heart of God Himself. To those who came after a season of loss, like ,my wife Rayanne did for me your love is a testimony that healing is still possible, that hope is still real, and that God is still writing beautiful stories in the ruins.

Chapter 12

1 Samuel

Our Anointed King and Voice of Silence

Chapter 12. 1st Samuel

Our Anointed King and Voice of Silence

"The LORD does not look at the things people look at. People look at the outward appearance, but the LORD looks at the heart."

— 1 Samuel 16:7

1st Samuel begins in barrenness and ends in battle. It introduces us to a broken priesthood, a praying mother, a rising prophet, a rejected king, and a shepherd boy no one saw coming. Through it all, God is preparing a King after His own heart, and pointing us to the true King to come.

When God Seems Silent

It all starts with Hannah, a woman whose womb is closed but whose spirit is open. She prays in anguish, so deeply moved that the priest assumes she's drunk, but she's not drunk, she's desperate. Her desperate prayer becomes the birth of Samuel, the last judge and the first prophet of a new era.

"In those days the word of the Lord was rare; there were not many visions." — 1 Samuel 3:1

Sound familiar? Ever been in a season when God's voice seemed rare? When prayers felt like they hit the ceiling? When the only thing louder than your trauma was the silence? Then suddenly, He speaks.

"Samuel! Samuel!" — 1 Samuel 3:10

God's voice breaks through the silence not to the elite, not to the powerful, but to a child sleeping near the Ark. That's the kind of King we follow. One who whispers to the overlooked.

A Misplaced Crown

Israel demands a king to be like the other nations. God gives them Saul a tall, handsome, powerful man who starts well but ends tragically. His fall reminds us external strength means nothing without internal surrender.

"To obey is better than sacrifice." — 1 Samuel 15:22

Saul lost the throne because he disobeyed and tried to cover it with religious excuses. We do the same when we try to act holy while ignoring the voice of the Lord.

But where Saul failed, Jesus succeeded. He obeyed fully. He didn't grasp for a throne He came to serve. He didn't hide behind religion He lived in perfect communion with the Father.

"Your attitude should be the same as that of Christ Jesus… He humbled Himself and became obedient to death, even death on a cross." — Philippians 2:5–8

David: A Shadow of the Coming King

When Samuel anoints David, nobody expected it. He's the youngest. The shepherd. The afterthought. Even his own father didn't invite him to the sacrifice.

"Are these all the sons you have?" — 1 Samuel 16:11

David is chosen not because of appearance but because of his heart. He becomes a warrior, poet, and king, but he also becomes a symbol of the Messiah to come.

David defeats Goliath with faith and a stone. Jesus defeats sin and death with a cross and an empty tomb. David leads Israel into battle. Jesus leads us into victory. David is flawed but chosen. Jesus is flawless and still chooses us.

Anointed in the Waiting

I've felt like David more than once, not the crowned king everyone celebrates, but the kid in the field nobody thought would make it. I was written off early by family, by foster homes, by the system. Fourteen homes before I was 9 years old. Each move another reminder that I didn't belong anywhere. I wandered, rebelled, enlisted, and thought the uniform would finally give me an identity.

But even in the Army, even downrange with the 173rd Airborne in Afghanistan, I was still carrying wounds inside I didn't know how to name. I came home to more battles divorce, betrayal, addiction, guilt, anger, and numbness. I was a soldier on paper, but inside, I was still a kid in the field, trying to figure out who I was.

Yet, through it all, God had an anointing on my life even when I didn't believe it, even when I ran from it, even when I drowned it in distraction and self-destruction. He saw me when I felt invisible. He knew the Goliaths I had faced both overseas and in my own heart and He knew the ones still ahead.

You might feel that way too, forgotten, overlooked, left in the pasture while everyone else gets chosen. Maybe you're still fighting battles no one sees. Maybe you think your chance to step into your calling is gone. Here's the truth: God doesn't forget who He anoints. His plans for you are not cancelled by your past, delayed by your detours, or undone by your mistakes.

David's anointing didn't mean instant promotion it meant preparation. God was shaping him in the field, in the hidden places, in the seasons of waiting and if you're in that place right now, know this waiting is not wasted when God is in it.

The same God who rescued David "from the paw of the lion and the paw of the bear" is the One who will rescue you from the hand of whatever stands in your way be it trauma, shame, addiction, or fear.

You are not forgotten. You are not disqualified. You are anointed in the waiting, and when the time is right, God will call you forward. Until then keep tending the sheep. Keep trusting the Shepherd. The battle may be bigger than you, but the victory has already been won for you.

"The LORD who rescued me from the paw of the lion and the paw of the bear will rescue me from the hand of this Philistine."

— 1 Samuel 17:37

Chapter 13

2 Samuel

Christ the King of Covenant and Compassion

Chapter 13. 2nd Samuel

Christ the King of Covenant and Compassion

"Your house and your kingdom will endure forever before me; your throne will be established forever." — 2 Samuel 7:16

If 1st Samuel is about finding the king, 2nd Samuel is about what kind of king we truly need. David finally takes the throne. The shepherd becomes sovereign. The man after God's heart becomes a ruler after God's promise. But the rise to glory doesn't come without failure, bloodshed, betrayal, and devastating consequences.

David's story is beautiful and brutal. Just like real life. Just like many of ours, but even in the wreckage, God's covenant never breaks. That's where we meet Christ in 2nd Samuel not just in the victories, but in the valleys. Not just in David's kingship, but in the covenant that would lead to Jesus, the King of Kings.

The Covenant King

In chapter 7, David wants to build a house for God. But God turns the tables:

"The LORD declares to you that the LORD Himself will establish a house for you… I will raise up your offspring… and I will establish the throne of his kingdom forever."

— 2 Samuel 7:11–13

This promise isn't ultimately about Solomon—it's about Christ.

"The angel said to her, 'The Lord God will give Him the throne of His father David… and His kingdom will never end." — Luke 1:32–33

Jesus is the true Son of David. Not just a king, but a covenant keeping King. Not just anointed, but eternal. Not just victorious in battle, but victorious over sin, death, and hell.

David's Fall and Christ's Faithfulness

Then comes David's darkest chapter. Bathsheba. Adultery. Deception. Murder.

He had the power to take whatever he wanted and he did, but the consequences were devastating. It's a painful reminder: Even the best leaders are broken.

"Against You, You only, have I sinned and done what is evil in Your sight…" — Psalm 51:4 (David's confession)

But God doesn't cancel the covenant.

Yes, David repents. Yes, he faces loss. But God's promise stands because the covenant was never based on David's perfection, but on God's mercy. Jesus is the anti-David in this moment. Where David saw and took, Jesus saw our sin and gave Himself instead. Where David abused power, Jesus laid His power down.

Mercy in the Midst of Collapse

Even in David's unraveling family his sons' rebellion, his grief, and his heartbreak we see a shadow of the Gospel.

When Absalom dies, David weeps: "O Absalom, my son… would I had died instead of you!"

At the Cross, Jesus does die instead of us.

"God demonstrates His own love for us in this: While we were still sinners, Christ died for us." — Romans 5:8

That's the difference between human kings and the King of Heaven.

Broken Kings and a Better One

David's story hits close for me. I've known leadership and failure. I've stood on mountain tops and sat in ashes. I've tasted victory and collapsed under the weight of my own choices.

There was a time I thought God was done with me. When the weight of trauma, addiction, betrayal, or guilt made me feel disqualified. David's story reminds me and should remind you that failure is not the end of the covenant.

God isn't surprised by our collapse. He's already made a way through it. Jesus sits on David's throne not because David was perfect, but because grace is.

You don't need to be a flawless warrior. You need to be a faithful one. You need to keep coming back to the King who never fails.

Chapter 14
A Story of Deliverance

The Christ Who Breaks the Chains of Addiction

Chapter 14. A Story of Deliverance

The Christ Who Breaks the Chains of Addiction

"It is for freedom that Christ has set us free. Stand firm, then, and do not let yourselves be burdened again by a yoke of slavery." — Galatians 5:1

The First Hit

I took my first hit of weed when I was eight years old.

I didn't even understand what it was fully but it did something. It numbed. It distracted. It created a fog that made the chaos around me feel manageable. What started as a childhood escape became a stronghold that lasted nearly three decades.

The Smoke Screen

Weed was my first escape. I was only eight years old the first time I got high. I was just a kid searching for something to quiet the noise, something to belong to, something to feel like control. What started as rebellion turned into rhythm. By the time I was a teenager, it was a lifestyle. After the war, it wasn't about fun anymore it was survival. I used it to cope with the flashbacks, the anger, the guilt, the shame and countless sleepless nights. Every time I said, "This is the last time," I'd circle back. I used it to manage my trauma, my anger, my shame. But instead of healing me, it hollowed me out. I was "functional" enough to get by. But deep down, I knew I was enslaved.

"You are a slave to whatever has mastered you." — 2 Peter 2:19

Beyond Weed: Other Chains I Wore

Weed wasn't my only battlefield it was just the beginning. Pain meds, alcohol, cocaine, adrenaline and porn all had their turn in my life. They promised relief but delivered chains.

Cocaine had its hooks in me. What started as a way to escape the chaos in my head became a chain I couldn't break. The high felt like power for a moment but it always came with a crash of shame, broken trust, and emptiness that grew louder each time. It never satisfied. It never healed. It never loved me back.

Pain meds started as a necessity. Used to help with lingering injuries, they were prescribed to help me function, and for a while, they did. But it wasn't just my body that hurt—it was my mind, my heart, my soul. The pills didn't just numb the physical pain they dulled everything. They made the darkness quieter. They made it easier to ignore what was really going on inside me. Before long, I wasn't taking them just to manage pain, I was taking them to manage life. To silence the memories. To push away the anxiety. To escape the war that never really ended. Like every other false savior, they took more than they gave.

Alcohol entered my life long before I understood its cost. I was just a kid when I started drinking too young to know I was trying to numb something, neglect, that had already started aching deep inside me. After deployment, it went from casual to constant. The bottle became a companion I trusted more than people. In the motorcycle club culture, drinking wasn't just accepted, it was expected. It gave me a false sense of control, a fake identity, and a temporary peace, but the more I drank, the more I lost who I was. Like weed, which had been with me since

I was eight, alcohol was another mask I wore to avoid the war within. It didn't heal my trauma, it buried it.

Pornography was the silent snare. It didn't leave track marks. It didn't make me slur my speech or fail a drug test. It did rewired my brain and robbed my heart. What started with curiosity as a young man turned into a coping mechanism after war, trauma, and loneliness set in. It became a place I went when I didn't want to feel, when I wanted control, escape, or comfort without the risk of intimacy. It promised pleasure, but it delivered shame. I carried it like a secret wound, thinking no one else struggled like I did.

Each addiction whispered, "This will make you feel better," but all of them led me to the same empty place alone, guilty, and more broken than before.

Jesus didn't just call me out, He called me back. Back to life. Back to truth. Back to the man He created me to be. In the places where cocaine, alcohol, weed, pills, and porn all promised peace, He gave me Himself. He stepped into every chain I wore and shattered them, not because I was strong, but because He is. He didn't wait for me to clean myself up; He walked right into the mess with mercy in His hands. What each addiction falsely promised comfort, control, escape, and worth, Christ fulfilled with His presence. He didn't just forgive me, He healed me. He restored my mind. He awakened my soul, and if He did it for me, He can do it for you. You are not beyond saving. You are not too broken. The story isn't over. In every book of Scripture, from Genesis to Revelation, the message is the same: Christ is still the hope for the addicted, the ashamed, and the shattered. His rescue mission is already underway.

The Veteran Battle with Addiction

Many veterans face this fight. War doesn't always stay overseas it follows us home. It lingers in the silence after the explosions, in the guilt we carry, in the memories that won't fade, and in the pain we can't explain. Addiction can become a way to function, to sleep, to feel something, or to feel nothing, but it's not the way forward. It's a trap disguised as relief.

Alcohol is one of the most common addictions among veterans, often used to numb PTS, fight off insomnia, or fill the void of isolation, but the bottle doesn't heal it only deepens the wound.

Opioids and prescription meds are regularly misused after injuries, surgeries, or chronic pain. They start in a bottle marked "relief" but often lead to stronger chains like heroin and fentanyl.

Cocaine may seem like an upper that gives energy, focus, or control but it leaves you spiraling in shame, paranoia, and dependence. What feels powerful for a moment leaves you powerless in the end.

Marijuana legal in many states can become a psychological and spiritual crutch. It may feel like it helps you cope, but it often keeps you stuck, numbing pain that Christ wants to heal.

Pornography quietly enslaves countless service members and veterans. It offers false comfort while feeding guilt, secrecy, and broken relationships. It damages how we see others and how we see ourselves.

Gambling, adrenaline addiction, and rage are other less spoken chains in the veteran community. The rush becomes a drug, and eventually, it robs us of peace, connection, and even safety.

If you're reading this and recognizing yourself, you are not alone, and you are not hopeless. You're not too broken for God.

These chains may be heavy, but they are not unbreakable. Jesus came to set captives free. He stepped into the fire, into the shame, into the battlefield of the soul and He hasn't stopped rescuing people like us. Your healing is not behind you, it's ahead. One step at a time. One surrender at a time. One page of Scripture at a time.

Trying to Quit and Failing.

There were times I tried to quit cold turkey, halfhearted attempts, or motivated by fear after a close call. I'd quit for a few weeks, then relapse. I'd make promises to myself and others, then break them. I'd pray and still feel powerless. The truth? Addiction isn't just a bad habit. It's a spiritual battle and I was trying to fight it in my own strength.

The Breaking Point

By the time I turned 36, the fog wasn't just emotional it was spiritual. I knew I was saved. I had seen God work in other areas of my life, but I still carried this chain. Then one day it snapped. Not because I got stronger, because God delivered me. As I was driving that still small voice told me "I'm all you need", It was so loud and clear I could swear Jesus was sitting right next to me riding shotgun, and even though He was not there physically or the voice was not audible He was for sure right there with me, and has never left my side.

I can't fully explain it, but I can tell you the truth, Since that day over three years ago, I have had zero cravings, zero thoughts, zero desire to go back to weed. It's not because I'm strong. It's because Christ is real, and His freedom is real.

Christ, the ChainBreaker

"So, if the Son sets you free, you will be free indeed."

— John 8:36

That's not a metaphor. That's not just a hopeful slogan. That's a promise. Jesus didn't just come to forgive our sins He came to set us free from every addiction, every idol, and every escape route that holds us back from living fully in Him.

I smoked for almost 30 years. I drank to numb. I took pills to escape. I put more powder up my nose than I can remember. I looked at things that cheapened intimacy. Now I walk clean, not because I fought my way out, but because I surrendered to the One who could. That's not to say I don't still struggle because I do, but I believe that just like He has delivered me from so much he can continue to delivery me for every single chain that tries to hold me down.

What Are You Still Holding?

Maybe it's not weed for you. Maybe it's alcohol, porn, pills, anger, control, or bitterness. Maybe it's something people applaud you for but you know it's killing you inside.

Christ is strong enough to break it. Not just someday. Not just in heaven. Now! It may take time. It may take counseling. It may take community, confession, and learning new patterns, but it always starts with this, Surrender. When freedom comes you'll know. Because you won't be white knuckling sobriety. You'll be breathing in grace.

Help for the Veteran in the Fight

If you're struggling, don't fight alone. Help is real, available, and closer than you think:

Veterans Crisis Line – Call or text 988, then press 1 or visit www.veteranscrisisline.net – 24/7 confidential support for veterans and their families.

Mighty Oaks Warrior Programs Faith based, peer to peer recovery and resilience programs for veterans, active duty, and spouses. www.mightyoaksprograms.org

Celebrate Recovery – www.celebraterecovery.com – Christ centered recovery program for all hurts, habits, and hangups.

Alcoholics Anonymous (AA) – www.aa.org – Support groups and resources for those battling alcohol addiction.

Narcotics Anonymous (NA) – www.na.org – Global support network for anyone struggling with drug addiction.

Covenant Eyes / Accountable2You www.covenanteyes.com – Internet accountability tools to fight pornography.

Veterans Affairs Substance Use Disorder Treatment – www.mentalhealth.va.gov/substanceabuse.asp–VA resources for inpatient and outpatient addiction treatment.

You may think you've tried everything—but don't give up before you've tried surrender.

Chapter 15 1st Kings

Christ the Wisdom and Glory Greater Than Solomon

Chapter 15. 1st Kings

Christ the Wisdom and Glory Greater Than Solomon

"But will God really dwell on earth? The heavens, even the highest heaven, cannot contain You. How much less this temple I have built!" — 1 Kings 8:27

A Kingdom of Brilliance and Breakdown

The book of 1st Kings is a story of both glory and collapse. It begins with the brilliance of Solomon's kingdom gold, peace, power, wisdom, and a temple so magnificent that the priests could not even stand when God's presence filled it. The nation enjoyed stability, prosperity, and influence that the world around them could not ignore.

From the outside, it looked unshakable. But as any veteran knows, appearances can be deceiving. Armies can look strong until morale breaks. Leaders can seem wise until their judgment falters. A kingdom can appear secure while cracks form beneath the surface. In 1 Kings, those cracks begin early and widen quickly. The kingdom that began with unity and worship soon drifted toward compromise and idolatry.

Christ is still present not in the politics, but in the promise. Not in Solomon's throne, but in the temple he built. Not only in Elijah's fire, but in the whisper that follows. Even in the decline, God is working, calling His people back, raising up prophets, and keeping His covenant alive.

The Rise of Solomon: Wisdom from Above

When Solomon begins his reign, he is humble: "I am only a little child and do not know how to carry out my duties." — 1 Kings 3:7

He is young, inexperienced, and aware of the weight of leadership. God offers him anything riches, victory, long life and Solomon asks for wisdom. God grants it, and the kingdom flourishes. Nations come to hear his counsel. Trade routes thrive. Peace reigns. Yet Jesus would one day say of Himself: "Something greater than Solomon is here." — Matthew 12:42

Solomon was wise, but Christ is Wisdom incarnate (1 Corinthians 1:24). Solomon ruled with splendor, but Christ rules with grace and truth. Solomon built the temple, but Christ is the temple: "Destroy this temple, and I will raise it again in three days." — John 2:19

The Temple: Glory in the Middle of the People

Solomon's temple was breathtaking crafted with gold, cedar, and onyx. Every measurement, every material, every detail was intentional. During its construction, no sound of hammer or chisel was heard (1 Kings 6:7) a symbol of reverence and peace. The people waited, watched, and wondered if the glory of God would come.

When it was completed, Solomon prayed, and the glory of God filled the temple like a cloud. "When the priests withdrew…the glory of the LORD filled the temple." — 1 Kings 8:10–11 God had come to dwell among His people. But this was only a shadow of what was to come. That glory would one day depart. The temple would fall. Because what they truly needed was not a building, but a Person who could dwell with them forever.

Elijah: Fire from Heaven, Whisper on Earth

The second half of 1st Kings introduces Elijah, one of the boldest prophets in all of Scripture. He calls down fire on Mount Carmel. He confronts corrupt kings and false prophets. He prays, and the rain stops for three years. He even outruns a chariot on foot. Yet when Jezebel threatens his life, Elijah breaks. He runs into the wilderness and prays, "I have had enough, Lord… take my life." — 1 Kings 19:4

That verse hits hard, especially for those who have faced depression, suicidal thoughts, or felt like their mission had outlasted their strength. God's response is tender. He does not rebuke Elijah. He sends food. He provides rest. He listens. Then He speaks not in the wind, not in the earthquake, not in the fire, but in a still small voice.

That is Christ. He is the whisper that calls you back from the edge, the gentle presence that meets you in the cave, the Commander who will not leave a soldier behind.

Greater Than the Gold

I used to think I had to get my life perfectly together before God would show up, as if I needed to build my own "temple," polished and impressive, before He would fill it. In the military, we were trained to make things look squared away uniforms crisp, gear inspected, and performance flawless. On the outside, you could be the picture of strength while inside you were falling apart. I carried that same mindset into my faith.

1 Kings reminds me: even the most glorious temples fall. Even the wisest leaders can drift. Solomon's wisdom couldn't keep him from idolatry. His gold covered temple couldn't hold the glory forever, and Elijah fresh off one of the greatest spiritual victories in Scripture found himself ready to quit, hiding in the wilderness.

Veterans know that pattern. You can go from the high of a successful mission to the crushing weight of life after service. You can feel the "glory days" slipping away and wonder if God still sees you. You can go from leading the charge to wanting to disappear, and like Elijah, you might think the fight has taken too much out of you to keep going.

The promise of God still stands because it's never been about your ability to hold everything together. It's about His ability to hold you. Jesus doesn't dwell in buildings made by human hands. He doesn't wait for you to put your armor back on or fix the cracks in your walls. He dwells in broken people made whole by grace.

You don't need a palace. You need a Savior. You don't need to rebuild what was lost. You need to rest in the One who cannot be lost. The gold fades. The glory changes. The fire eventually burns out, but the whisper, the voice of Christ remains, and when you hear it, you know you're not done yet.

Chapter 16

2nd Kings

Christ the Faithful Word When Kingdoms Fall

Chapter 16. 2nd Kings

Christ the Faithful Word When Kingdoms Fall

"The word of the LORD spoken through His servants the prophets was fulfilled." — 2 Kings 24:2

The book of 2nd Kings reads like a slow motion collapse.

The cracks start small another king drifts from God, another compromise with idols, and another prophet ignored. Over time, the cracks widen. The northern kingdom of Israel falls to Assyria. The southern kingdom of Judah holds out a little longer, but eventually, it too is taken into exile. The temple once filled with God's glory is burned to the ground. The people are scattered, the royal line appears broken, and the promises of God seem like a distant echo.

That's the thing about collapse, it's often not sudden. It's slow, quiet, and almost invisible until the walls are already caving in.

For many veterans, that's exactly how the aftermath of war works. The real collapse doesn't always happen on the battlefield it happens at home. It starts with sleepless nights. Maybe an extra drink to calm down. A prescription that turns into a dependency. A little more isolation. Fewer calls to friends. Fewer prayers. The uniform comes off, but the weight never leaves, and one day, you realize the "temple" of your life, the things you thought would always stand, your marriage, your health, your sense of purpose are crumbling.

PTS doesn't always kick in the door it seeps in like water, wearing you down. Substance abuse doesn't always start with a conscious decision it creeps in through pain, boredom, or the need to forget. Guilt doesn't always shout it whispers until you believe it. Like Israel's kings, you think you can manage it, negotiate with it, keep it under control, but idols don't play fair they always take more than they give.

Yet through all of this 2nd Kings gives us a lifeline:

The Word of the Lord does not fail. Even as Israel and Judah spiral, God keeps speaking. Prophet after prophet is sent. Warnings are given. Mercy is offered. The people might ignore it, but the Word stands. That's where we find Christ in 2nd Kings not sitting on a throne in Jerusalem, but in the unshakable promises of God, and in the faithful remnant who refuse to bow.

From Elijah to Elisha: A Double Portion

The book opens with a dramatic handoff. Elijah the fiery prophet who called down heaven's flames is taken up in a whirlwind. Elisha, his student, refuses to leave his side until the very end. When Elijah asks what he wants, Elisha replies, "Let me inherit a double portion of your spirit." (2 Kings 2:9)

God grants it. Elisha's ministry becomes a reflection of Christ's healing the sick, multiplying food, cleansing the unclean, and even raising the dead. He takes the mission further, showing that God's power doesn't end with one man it continues, multiplies, and points to the One who would come with the Spirit and without measure.

Kings Fall, but God Does Not

The kings in 2nd Kings come and go some try to reform, but most fail. The nation keeps chasing idols. Judgment finally comes, but even then, God is still faithful. He keeps a remnant. He keeps the promise alive.

Jesus in 2nd Kings: The Final Prophet and True Temple

Elisha multiplies bread, Jesus feeds thousands. Elisha heals a leper; Jesus heals many and welcomes outsiders. Elisha raises the dead; Jesus is the Resurrection and the Life. The temple falls: Jesus becomes the Temple that no power can destroy.

When the veteran's life feels like 2nd Kings when the collapse feels inevitable, when guilt feels permanent, when the "temple" is in ashes the story still isn't over. The same God who carried His people through exile can carry you through addiction, PTS, and guilt. The collapse isn't the end. It's the ground where God begins rebuilding.

When the Walls Are Falling

I've lived my own version of 2nd Kings. It didn't happen overnight it was a slow fade. A drink here. A hit there. A sleepless night that turned into months of insomnia. The war was over, but the battle in my head never stopped. I told myself I was "managing," but really, I was crumbling.

Like the kings of Israel and Judah, I thought I could make compromises with my idols weed, alcohol, pills, porn, and anger and still keep the kingdom standing. I thought I could outrun guilt and outdrink shame, but the thing about idols is they never stay in their place. They demand more and more until they own you completely.

If you've been there or if you're there right now you know how it feels when the collapse begins: The relationships you thought would never break start fracturing. The sense of purpose you used to live by fades into apathy. The temple of your life, the place where you thought God would always dwell, feels empty.

Here's what 2nd Kings and my own life have taught me:

Even when the kingdom falls, the Word of the Lord does not fail. Christ is not afraid of your ruins. He isn't waiting for you to rebuild the temple before He moves back in. He comes into the rubble. He speaks into the silence. He calls you out of the ashes and back into His presence.

You may feel like you're in exile, cut off, and alone, far from God's favor but exile was never the end of Israel's story, and it's not the end of yours.

The same God who kept His promises through centuries of rebellion will keep His promises to you. The same Christ who raised the temple of His own body from the dead can raise your life from the dust.

Veteran, your collapse is not your conclusion. PTS is not your identity. Addiction is not your forever. Guilt does not have the final word. The final word belongs to the One who said, "It is finished."

Chapter 17

1st & 2nd Chronicles

Christ the Better King, the True Temple, and the Fire That Fills

Chapter 17. 1st & 2nd

Chronicles Christ the Better King, the True Temple, and the Fire That Fills

"Yours, LORD, is the greatness and the power and the glory and the majesty and the splendor… In Your hands are strength and power to exalt and give strength to all."

— 1 Chronicles 29:11–12

When most people think of 1st and 2nd Chronicles, they think of a lot of names, repeated history, and temple details, but these books are far more than historical records they are the heartbeat of a people remembering who they are, even after everything has fallen apart.

1st Chronicles opens with nine chapters of genealogy, name after name, generation after generation. We might be tempted to skip it, but God doesn't waste words. Those names remind us that God doesn't forget His people. Even after exile, even after the land, the throne, and the temple were lost, the line leading to Christ was still intact.

From Adam to Abraham, from Judah to David, from David to the returning exiles, every name said, The promise still stands. That promise was pointing to the King who would never fail, Jesus Christ.

Veterans: Remembering Who You Are

When I read those long lists of names, I think about my brothers and sisters in the military, those still serving, those who came home, those still fighting wars in their minds, and those who never made it home. In the same way the exiles needed to be reminded of their lineage, veterans need to be reminded of theirs, not just the military branch or the unit patch, but the deeper identity we have in Christ.

Combat can make you forget who you are. PTS can make you think you'll never be yourself again. Addiction, guilt, and survivor's remorse can whisper that you're just a shell of the soldier you were, but if you belong to Jesus, your identity isn't in what you did, what you've lost, or what you've survived. It's in whose you are.

Israel needed to remember they were still God's chosen people, even in exile. Veterans need to remember we are still God's beloved sons and daughters, even in our brokenness. The war may have changed you, but it did not erase you. Your name is still known in heaven. Your life is still written into God's story.

David: The Warrior Who Prepared for Peace

Unlike Samuel and Kings, 1st Chronicles retells David's life through a lens of worship and preparation. His failures are not the focus. Instead, we see his deep desire for God's glory. He's told he cannot build the temple because of the blood on his hands, but he doesn't sulk or retreat. He gathers gold, organizes priests and musicians, writes psalms, and casts vision for Solomon to build it.

"The task is great, because this palatial structure is not for man but for the Lord God." — 1 Chronicles 29:1

David knew it wasn't about a building it was about the presence of God in the middle of His people.

Worship in the Wilderness

Chronicles devotes massive space to worship leaders, gatekeepers, and musicians because in Israel's darkest seasons, worship anchored them. Even when the temple was gone, the memory of God's presence remained.

I've learned that same truth the hard way. I've worshiped in the good times and the bad, in seasons of joy and prosperity and in seasons of addiction, depression, and silence. The place may change. The season may change, but Christ's worthiness never does.

The Temple—and the One Who Would Replace It

When we step into 2nd Chronicles, Solomon builds the temple his father prepared for. The dedication is breathtaking gold, cedar, carved cherubim, and an altar ready for sacrifice, but what made it holy wasn't the craftsmanship. It was the presence of God.

"Fire came down from heaven… and the glory of the LORD filled the temple." — 2 Chronicles 7:1

The priests couldn't even stand under the weight of that glory. Here's the thing: that temple wouldn't last forever. It would be looted, burned, and leveled. Because God's ultimate plan was never to live in a building, He wanted to live in us.

"Don't you know that you yourselves are God's temple and that God's Spirit dwells in your midst?" — 1 Corinthians 3:16

Jesus is the true Temple. He is the glory that fills the empty. He is the fire that never burns out.

From Revival to Ruin

Chronicles gives us the stories of kings, some faithful, some foolish. There are seasons of revival when the people humble themselves, and seasons of ruin when they turn to idols. Through it all, God remains faithful.

I feel this as a veteran and former addict. I've seen good things collapse marriages, faith, sobriety, and purpose. I've watched my own "temple" fall to pieces, and yet, God didn't abandon me. He called me back. He rebuilt me from the ruins.

The turning point wasn't on a battlefield or in a high it was on my knees, crying out the words of King Jehoshaphat:

"We do not know what to do, but our eyes are on You." — 2 Chronicles 20:12

And just like He did for Judah, He answered not just removing what enslaved me, but filling the empty place with Himself.

The God Who Rebuilds Temples

The message of Chronicles is simple but powerful God's promises outlast our collapses. You might feel like your "temple" has burned down: You had faith once, but it's cold now. You had sobriety, but you slipped. You had purpose, but you feel lost. You had a marriage, but it's hanging by a thread.

You're not the first to feel that way, and you won't be the last. Israel saw their temple destroyed and their people carried off, but God's Word still came true. The line to Christ was unbroken. The best part is that when He rebuilds, He doesn't just restore what was lost, He fills it with a greater glory.

"The glory of this present house will be greater than the glory of the former house." — Haggai 2:9

That means your story isn't finished. The God who kept the line unbroken through centuries of failure, exile, and loss is the same God who can keep you. You are not forgotten. You are not too far gone. You are not too broken for Him to fill you with His presence again.

Chapter 18

Ezra

Christ the Foundation We Return to

Chapter 18. Ezra

Christ the Foundation We Return to

"Though we are slaves, our God has not forsaken us in our bondage. He has shown us kindness… to grant us new life to rebuild the house of our God and repair its ruins…"

— Ezra 9:9

Ezra is a book for the comeback story.

After seventy years in exile, a remnant of Israel returns to Jerusalem not to reclaim political power or military dominance, but to rebuild the altar and restore worship.

They weren't the same people who left. They were children of trauma Born in Babylon, Used to loss, Haunted by judgment, Carrying the stories of their parents' failure, and yet, God brought them back, not because they were strong, but because He is faithful.

The Altar First

When the remnant returned to Jerusalem, they didn't start with city walls or homes. They didn't strategize military defense.

"Despite their fear… they built the altar on its foundation." — Ezra 3:3

The altar came first because they understood something we often forget nothing matters without worship. No wall can protect a people who have forgotten how to bow. No kingdom can rise without sacrifice at its center. No rebuilding is complete without God's presence at the foundation. That truth still holds.

As a combat veteran and recovering addict, I've rebuilt many parts of my life: Mental and emotional discipline, marriage and trust, career and calling, and faith and ministry. None of those could stand without Jesus at the center. He is my altar. He is where I lay down my pride, trauma, temptation, and shame.

Veterans and the Call to Come Home

I've seen soldiers try to rebuild life after deployment without touching the altar, without giving their pain to Christ.

It doesn't work. We patch the walls, but the cracks keep showing. We clean the surface, but the rot stays underneath.

Coming home from war can feel like exile in reverse. Everyone else kept living while you were gone. You're back in familiar streets, but you feel like a foreigner. Your body's home, but your soul is still out there.

That's when the temptation comes: To isolate. To self-medicate. To believe you're too far gone to be restored. Ezra's story reminds us, we can come home. Not to what we were before, but to something new.

"Return to Me," declares the LORD Almighty, "and I will return to you." — Zechariah 1:3

Your identity isn't in the uniform, the discharge papers, or the scars. It's in the One who calls you by name and invites you to build your life on Him again.

Christ in Ezra: The Altar, the Priest, the Word

Ezra points us to Jesus at every turn:

The rebuilt altar points to the Cross where the ultimate sacrifice was made. The restored priesthood foreshadows Christ, our

Great High Priest. The public reading of the law anticipates Christ, the Word made flesh. The repentance of the people shows us God's heart for the contrite.

Ezra isn't just about coming back to a place. It's about coming back to a Person. It's about rebuilding around Jesus, the unshakable foundation.

When You've Been in Babylon Too Long

I've lived in Babylon literally in war zones and spiritually in the wastelands of addiction. I've been in exile from peace, from joy, from prayer, from hope. I've heard the lie: "You'll never get back to where you were." The truth is, God doesn't bring you back to what you were before. He brings you forward into something better.

"See, I am doing a new thing! Now it springs up; do you not perceive it?" — Isaiah 43:19

He uses the scars. He uses the time you think was wasted. He uses the wreckage to rebuild something sacred.

Start With the Altar

If you're trying to put your life back together after a relapse, divorce, depression, or betrayal I urge you to start with the altar.

Not with perfection. Not with performance. Not with a plan to "do better this time." Start with Jesus. Lay it all down, again.

The guilt. The fear. The pride. The excuses. The bitterness. The numbness. Ezra's people didn't have every detail sorted. They came home and lit the fire again. That was enough for God to meet them there. It's enough for Him to meet you, too.

"Come to Me, all you who are weary and burdened, and I will give you rest." — Matthew 11:28

Chapter 19
Nehemiah

Christ the Restorer of Broken Walls and Wounded People

Chapter 19. Nehemiah

Christ the Restorer of Broken Walls and Wounded People

The book of Nehemiah is more than an account of ancient construction. It's the story of a man who heard of his people's ruin, let the news break his heart, and allowed God to turn his grief into action. Nehemiah wasn't a priest, prophet, or soldier—he was a cupbearer. Yet he carried a burden big enough to move a nation.

When Nehemiah heard Jerusalem's walls lay in ruins, he didn't shrug it off or wait for someone else to care. He wept. He fasted. He prayed. Then, he acted. In that movement, we see the heart of Christ, the One who left heaven's throne to enter the wreckage of our lives and rebuild what sin and suffering had destroyed.

Rebuilding with Resistance

The moment Nehemiah began rebuilding, opposition rose. Sanballat, Tobiah, and others mocked, threatened, and plotted to stop the work. Their weapons weren't just physical, they were psychological. They tried distraction, intimidation, and lies.

Nehemiah's answer was unwavering:

"I am doing a great work and I cannot come down." — Nehemiah 6:3

That verse is a lifeline when discouragement tries to pull you off the wall. I've heard the enemy's whisper: "You're not worth this. You'll never change. This work is too hard." But like Nehemiah, I've had to stand my ground, not because I'm strong, but because the One who called me is faithful.

As Rick Warren says in *The Purpose Driven Life*, "It's not about you." The mission God gives us is bigger than our comfort. It's His work, His glory, and His story unfolding.

Restoration of the Wall — and the People

In Wounded Warrior, Wounded Home, the truth is clear: the visible damage, broken walls, shattered routines, is only part of the problem. The deeper wounds are often invisible: fractured trust, crushed identity, and the weight of shame.

Nehemiah understood this. Rebuilding Jerusalem's wall wasn't just about security, it was about restoring dignity. In ancient cities, broken walls meant vulnerability and disgrace. Restoring them was a public declaration: "We are God's people again. We are not abandoned."

That's the kind of restoration Jesus brings. He doesn't just patch over the damage; He rebuilds your defenses, redefines your identity, and restores your sense of belonging. John Eldredge writes in Resilient that "God is not just trying to get you through; He's trying to restore your soul."

The wall was rebuilt in 52 days, but the people's healing would take far longer. Trauma doesn't evaporate when the physical work is done. Like Pete Walker explains in Complex PTS, recovery means learning safety, trust, and self-worth all over again. The rebuilding was as much about their hearts as it was about their city.

My Wall Took Years

I know what it is to live without walls, no boundaries, no defense against temptation or despair. War stripped me bare.

Addiction left me exposed. My internal city was wide open to every attack.

Christ came into that wreckage—not to shame me, but to restore me. Slowly, through prayer, Scripture, community, and discipline, He began to rebuild. Like Craig Groeschel teaches in *The Power to Change*, the transformation came from small, consistent steps, laying one brick at a time.

I had to work with one hand and fight with the other. David Goggins calls this "callousing the mind," pushing through resistance until perseverance becomes part of you. Like Tim Grover writes in *Relentless*, there's a mindset that refuses to go back to what broke you. That's what staying on the wall looks like.

Goliath Must Fall

Louie Giglio reminds us in *Goliath Must Fall*: "Your giant is dead; you just have to walk in the victory that's already yours." For Nehemiah's people, the "giant" wasn't just the rubble, it was the fear, shame, and hopelessness that had settled in during exile.

For us, it may be addiction, depression, fear, shame, or bitterness. Those walls don't get rebuilt by sheer grit, they rise because Christ already defeated the enemy. We rebuild, not to earn victory, but to live in it.

A Community on Mission

Nehemiah didn't work alone. Every family had a section of the wall. Goldsmiths, perfumers, priests, each took responsibility for their part. This is the Church in action: wounded people working side by side, rebuilding together.

You don't have to be fully healed to help. You just have to show up. Bonhoeffer wrote in *The Cost of Discipleship*, "When Christ calls a man, he bids him come and die." Part of dying to self is choosing to keep building even when it hurts, knowing the wall you help raise may protect someone else.

Don't Come Down

If God has placed you on a wall, whether it's rebuilding your sobriety, marriage, prayer life, or ministry, stay there. The enemy will tempt you with distraction, fear, and compromise. But your work matters. Your healing matters. Your faithfulness matters.

Lay one more brick today. Pray one more prayer. Forgive one more time. Keep your eyes on the One who strengthens your hands for the work.

Chapter 20

Esther

Christ The Hidden Hand in the Middle of the Battle

Chapter 20. Esther

Christ the Hidden Hand in the Middle of the Battle

"If I perish, I perish." — Esther 4:16

The book of Esther is unlike any other in the Bible. There are no visions, no burning bushes, no thundering voice from heaven. In fact, God's name is not mentioned even once, and yet, His fingerprints are everywhere. They are woven into each twist of fate, each royal decree, and each turning point in the story.

Esther's journey begins quietly. A young Jewish woman is taken from obscurity and placed in a position of influence during one of the most dangerous moments in her people's history. She doesn't seek the spotlight. She doesn't ask for the mission, but when the threat comes, she steps forward. She risks everything without knowing how it will end.

That is faith in the shadows. That is trust when the ground under your feet is shaking.

God in the Shadows

Some of the darkest seasons of my life felt like Esther's story. Not in the details, but in the silence. There were stretches where I couldn't hear God clearly and I didn't know if He was even there. There were no miraculous signs just pain, survival, and the feeling of being in exile.

I thought He had left me when I was shuffled through 14 foster homes before I was out of the 2nd grade. I couldn't put down the weed even though I hated myself for it. I came home from Afghanistan and didn't recognize the man in the mirror. My prayers seemed to bounce off the ceiling. Like Esther's story, God was still writing mine, behind the scenes, in the silence, in the moments I thought were wasted.

Jesus in Esther's Story

You won't see the name "God" in Esther, but you will see His character, and you'll catch glimpses of Christ. Esther risks her life to enter the king's presence on behalf of her people. Jesus entered death itself on our behalf so we could live. Esther stands in the gap for those marked for destruction. Jesus became the gap, the only Way back to the Father. Esther's courage reverses a death sentence. Jesus' resurrection rewrites the entire story of humanity. Esther points to Jesus, not with fireworks or parting seas, but with quiet obedience that changes history.

For Such a Time as This

Mordecai's words to Esther still feel like a thunderclap:

"Who knows but that you have come to your royal position for such a time as this?" (Esther 4:14)

This was never about royalty, it was about responsibility. You may not have a crown or a throne, but if you've been through trauma, addiction, or war and you're still here, there's a reason.

You carry a story that someone else needs to hear. You have scars that can help heal someone else's wounds. Your survival is not an accident, it's an assignment.

If you're in a season where God feels absent, remember silence is not absence. Look for Him in the timing of events you didn't plan, the people He places in your path, the unexpected strength you find when you thought you had nothing left, and the closed doors that protect you as much as the open ones.

Sometimes the miracle is simply that you're still standing. Sometimes the victory is that you didn't quit, and sometimes God's greatest work is done in the dark, where faith is refined without applause.

Chapter 21
Job

Christ, the Redeemer Who Answers io the Silence

Chapter 21. Job

Christ, the Redeemer Who Answers in the Silence

"Though He slay me, yet will I hope in Him." — Job 13:15

The book of Job is one of Scripture's most unflinching portraits of human suffering. It doesn't soften the blow or offer quick fixes. Instead, it walks us into the ash heap, sits down beside us, and lets us feel the weight of loss. It is a book for anyone who's asked hard questions, wrestled in the dark, or survived seasons when God felt absent.

When Everything Collapses

Job's life unraveled in an instant. Children gone. Health shattered. Wealth erased. Reputation shredded. Friends turned accusers, and silence from heaven.

I've been there, not in Job's exact shoes, but in the same storm. I've looked at the wreckage of my own life, failed drug tests, lost jobs, estranged relationships and felt abandoned. I've come home from Afghanistan with a war in my head and silence in my prayers. I've sat in church surrounded by worship and wondered why God wasn't speaking to me.

Job's story doesn't try to explain that pain away. It validates it, and then it shows us that even in the deepest loss, there can be hope.

Holding on When You Have Nothing Left

What amazes me most about Job isn't his patience, it's his persistence. Job kept speaking to God even when God didn't answer, and he clung to a hope he could barely see:

"I know that my Redeemer lives, and that in the end He will stand on the earth." (Job 19:25)

That wasn't polished theology it was a desperate cry from a broken man. Yet, in those words, Job prophetically points to Jesus. As J.I. Packer reminds us in *Knowing God*, "true faith is not about having all the answers—it's about knowing the One who does."

Christ in Job's Story

Job longed for a mediator someone who could lay a hand on both God and man (Job 9:33). Jesus is that Mediator (1 Timothy 2:5). Job wanted someone to speak for him in heaven. Jesus is our Advocate (1 John 2:1). Job cried for justice and restoration, Jesus bore our injustice, healed our wounds, and clothed us in His righteousness. The cross answers Job's plea. The resurrection answers his silence.

Faith in the Silence

Silence is one of the hardest parts of faith. It's in the quiet that doubts echo loudest, but as Dallas Willard says in *Hearing God*, "God's silence is never abandonment, it's often an invitation to deeper trust." Job's trial teaches us that God may not always explain, but He will always reveal Himself.

When God finally speaks in chapter 38, He doesn't give Job the "why." He gives Job Himself, His presence, His voice, His majesty, and that changes everything.

Grace for the Wounded

For veterans, trauma survivors, and anyone whose soul feels war-torn, Job's journey is a roadmap for grace. Healing doesn't happen all at once. It's brick by brick rebuilding: prayer, boundaries, honesty, Scripture, and showing up even when it hurts. As John Eldredge writes in *Resilient*, "God is not in a hurry. He is growing you into someone unshakable."

God's grace doesn't just restore what you lost it transforms who you are. Job came out of the storm with deeper humility, sharper faith, and a more intimate knowledge of God.

Redemption in the End

In the final chapter, Job's fortunes are restored, not just materially, but relationally and spiritually. His story reminds us that God can take the ashes of our lives and make them fertile ground for something new. As A.W. Tozer writes in *The Crucified Life*, "surrender is the doorway to resurrection power."

Redemption isn't the absence of scars, it's the testimony that God is greater than what broke you. My own scars from war, addiction, and years of self-destruction, don't define me. Christ does and in Him, like Job, I have hope that the story isn't over.

Chapter 22
Psalms Part 1

Christ, the Song in Our Suffering

Chapter 22. Psalms: part 1

Christ, the Song in Our Suffering

"My God, my God, why have You forsaken me?" — Psalm 22:1

The book of Psalms is a battlefield. It's where wounds bleed out in poetry, where kings cry like children, and where warriors drop their armor to kneel in weakness. It's the honest soundtrack of God's people sung in caves, in courts, in prison cells, on mountaintops, and in valleys so deep the sunlight never touches the ground.

When I came to the Psalms, I didn't bring a neat theological outline. I came with my trauma. I came out of foster homes, warzones, addiction, and nights I didn't think I'd survive. PTS and relapse were my traveling companions, and in these pages, I found voices that knew exactly how I felt.

I found David once a shepherd boy, later a warrior king crying out like I did. I found Asaph admitting that he envied the wicked when life felt unfair. I found nameless psalmists raging, repenting, and reaching for God like a drowning man gasping for air, and in the middle of it all, I found Jesus.

Jesus Is the Cry in the Darkness

There's a reason Psalm 22 opens with:

"My God, my God, why have You forsaken me?" (Psalm 22:1)

Those same words would be torn from the mouth of Jesus a thousand years later, as He hung on the Cross:

"My God, my God, why have You forsaken me?" (Matthew 27:46)

Jesus didn't just quote that Psalm, He embodied it. He stepped into ultimate suffering, carrying not only our sins but the crushing silence of the Father's absence. The Psalm even sketches the scene at Calvary:

"They pierce my hands and my feet… They divide my garments among them and cast lots for my clothing." (Psalm 22:16, 18)

The Psalms teach us that Christ doesn't meet us at a safe distance, He meets us in the darkest places. He knows betrayal, mockery, abandonment, and pain that won't let you breathe. He knows what it feels like when your prayers hit the ceiling and fall back unanswered.

When Combat Doesn't Stay on the Battlefield

There are Psalms I've prayed word for word from my own foxholes:

"How long, O Lord? Will You forget me forever?" (Psalm 13:1)

"I am poured out like water… My strength is dried up like a potsherd…" (Psalm 22:14–15)

"Save me, O God, for the waters have come up to my neck." (Psalm 69:1)

These are not soft, pretty verses, they are combat prayers. I've prayed them with dirt under my nails from fighting demons all night. I've whispered them when I couldn't feel God at all. I've shouted them through tears, trying again to quit weed after failing a hundred times, and I've learned something: God doesn't flinch at raw honesty.

Like David in the cave, I've cried in the dark and found that God wasn't afraid of my mess. He met me in it.

Christ in the Psalms of Suffering

Psalm 22 — The Crucified Christ

Psalm 31 — "Into Your hands I commit my spirit" (Jesus on the Cross, Luke 23:46)

Psalm 41:9 — The betrayal of a close friend (Judas)

Psalm 69 — The Rejected Savior

Psalm 109 — The suffering servant surrounded by enemies

This isn't just prophecy, it's solidarity. The Son of God prayed the same prayers we do, but without sin. That means He can carry our shame, our fear, and our unanswered questions without condemn-nation.

A Place for Warriors and the Wounded

For veterans and trauma survivors, the Psalms are not simply ancient songs, they're a survival manual. They give you language when all you have is groaning. They remind you that warriors in God's kingdom don't always stand tall, sometimes they collapse at His feet and call it worship.

In Psalms, I learned I didn't have to fake strength to be faithful. I could admit I was weak, confused, even angry, and still belong to Him. That's grace. That's redemption in real time.

You're Allowed to Be Honest. The Psalms gave me permission to stop pretending. I used to think I had to clean myself up before I came to God, that my prayers needed polish and order, but the Psalms show us: You can rage and still be righteous. You can doubt and still be faithful. You can weep and still worship, and you can relapse and still return. As Psalm 51:17 says: "A broken and contrite heart, O God, You will not despise." So go ahead pray your ugliest prayer. God is listening, and Christ is already interceding for you.

Chapter 23
Psalms Part2

Christ, the Song and Shepherd of the Soul

Chapter 23. Psalms: part 2

Christ, the Song and Shepherd of the Soul

Psalm 23: A Veteran's Psalm

Psalm 23 is often read at funerals, but for veterans and other survivors of trauma it is a survival song. We and cling to it in Afghanistan. Recited it during relapses. Whispered it after panic attacks, or quoted it when sleep won't come and nightmares won't leave.

"The LORD is my shepherd, I lack nothing. He makes me lie down in green pastures, He leads me beside quiet waters, He refreshes my soul." — Psalm 23:1–3

Trauma had rewired my mind. Violence had trained my body to stay ready. Even at home, I never stopped scanning, but slowly, Christ my Shepherd led me beside quiet waters.

"Even though I walk through the valley of the shadow of death, I will fear no evil, for You are with me." — Psalm 23:4

I've walked that valley with fellow soldiers and buried too many but He was with me, and that changed everything.

Jesus in the Psalms

Psalms is the most quoted Old Testament book in the New Testament, and it reveals Jesus in nearly every chapter. Here's the fuller picture:

Psalm 2 — The Anointed King

"You are My Son; today I have become Your Father" (Psalm 2:7).

Fulfilled in Jesus at His baptism and resurrection (Matthew 3:17; Acts 13:33).

Psalm 8 — The Crowned Humanity

"You made them rulers over the works of Your hands" (Psalm 8:6).

Applied to Christ in Hebrews 2, showing Him as the perfect Man who regains what Adam lost.

Psalm 16 — The Resurrection Hope

"You will not abandon me to the realm of the dead" (Psalm 16:10).

Peter quotes this in Acts 2:25–32 to prove Jesus rose from the dead.

Psalm 22 — The Crucified Savior

Pierced hands and feet, divided garments, mocked and abandoned—every line fulfilled at Calvary.

Psalm 23 — The Good Shepherd

Jesus identifies Himself as the Shepherd who lays down His life for the sheep (John 10:11).

The Rhythm of Restoration

The second half of Psalms beats with the rhythm of recovery:

Weep — Pour out your heart.

Wait — Trust in the silence.

Worship — Praise in the pain.

Walk — Move forward in faith.

Lead — Guide others to the Shepherd.

As a veteran, I've lived every line. I've learned that the same God who met David in the caves meets us in VA waiting rooms, Celebrate Recovery circles, and midnight prayers when the rest of the house is asleep. His voice hasn't changed. His mercy hasn't run out.

Follow the Shepherd, Even with a Limp

You don't have to walk perfectly to follow Jesus. You can follow Him limping. You can follow Him with night sweats, shaky faith, and battle scars. You can follow Him in broken boots and hand-me-down Bibles. Just follow.

"Surely goodness and mercy will follow me all the days of my life…" — Psalm 23:6

The Shepherd doesn't just go ahead of you He follows you, too. Covering your rear. Carrying your wounds, and calling you forward.

Chapter 24

Proverbs

Christ, the Wisdom of God Who Heals Our Wayward Paths

Chapter 24. Proverbs

Christ, the Wisdom of God Who Heals Our Wayward Paths

"Trust in the LORD with all your heart and lean not on your own understanding; in all your ways submit to Him, and He will make your paths straight." — Proverbs 3:5–6

Wisdom isn't just about what you know, it's about who you become.

The book of Proverbs is often seen as a collection of sayings for the wise, but for people like me who grew up on the streets, got high before I even hit puberty, flunked drug tests, hurt people I loved, and got burned by my own decisions, it became something deeper.

It became an invitation. An invitation to slow down, to listen, to heal, to become someone new. Because wisdom isn't earned. It's revealed, and it's found in Christ.

When Street Smarts Fail

Growing up in Detroit's inner city, I learned survival early. Not wisdom, just survival. I learned to lie, to hide, and to hit first. I learned to mask pain with weed. I started smoking at eight years old, long before I understood what addiction was doing to me.

I thought I was grown before I even knew who I was. The world applauded toughness but never taught truth, and let me tell you toughness alone will take you places your soul can't survive. Proverbs cut through the noise. "There is a way that appears to be right, but in the end it leads to death." — Proverbs 14:12

That was me thinking my instincts could save me, but my instincts were shaped by pain, not wisdom. My path was straight toward destruction, addiction, brokenness, pride and trauma. Wisdom from God was the only way home.

Jesus: The Voice in the Streets

When Proverbs speaks of wisdom as a voice calling out in the streets (Proverbs 1), I picture Christ walking into the chaos of my life. Michael Todd once preached, "God doesn't just meet you in the church, He'll find you in the middle of your mess." That's Proverbs wisdom in real life.

In the Bible, wisdom is not just a concept it's a Person. "Christ, the power of God and the wisdom of God." — 1 Corinthians 1:24 "In Him are hidden all the treasures of wisdom and knowledge." — Colossians 2:3

Jesus is Wisdom calling to the foolish, the addicted, the wandering, the war weary. He doesn't just hand out life tips, He gives Himself.

Proverbs in the Foxholes of Life

During deployment, I didn't sit around reading poetry and sipping coffee. I was watching for IEDs, keeping my rifle ready, and making split second decisions that could mean life or death, but later when the adrenaline stopped and the nightmares started, Proverbs became my battle manual for the mind.

Some verses became like flashbangs waking me up from destructive thinking:

Proverbs 26:11 — "As a dog returns to its vomit, so fools repeat their folly." I kept going back to what poisoned me. Jesus broke that cycle.

Proverbs 4:23 — "Above all else, guard your heart, for everything you do flows from it." Recovery was not just about quitting, it was about rebuilding my heart.

Proverbs 15:1 — "A gentle answer turns away wrath, but a harsh word stirs up anger." Pastor Brad Powell once said, "Wisdom is knowing the right thing to do; godliness is actually doing it." I think I am still learning this some days.

Proverbs 9:10 — "The fear of the LORD is the beginning of wisdom." Bishop T.D. Jakes says, "If you fear God, you'll fear nothing else." That truth gave me courage when anxiety tried to own me.

Lessons from Other Veterans

I've sat in circles with other vets, Mighty Oaks retreats, Celebrate Recovery meetings, VA waiting rooms. Our stories are different, but the theme is the same: we've all tried leaning on our own understanding, and it landed us in dark places.

One brother told me, "In Afghanistan, I had a map for the terrain but back home, I had no map for my mind." Proverbs gave him a map for the moral terrain of civilian life. Another said, "Wisdom doesn't mean you never mess up, it means you learn faster and repent sooner."

We all learned that survival mode keeps you alive in combat but can kill your relationships, your faith, and your peace in peacetime. Wisdom is what bridges the gap.

The Lies We Used to Believe

In Lies Men Believe, Robert Wolgemuth writes about the subtle deceptions men buy into, like "I can handle it," "I don't need help," or "Real men don't cry." I believed all three.

Proverbs exposes these lies and replaces them with truth:

Lie: I can handle it. Truth: "In all your ways submit to Him…" (Proverbs 3:6) Lie: I don't need help. Truth: "Plans fail for lack of counsel, but with many advisers they succeed." (Proverbs 15:22) Lie: Real men don't cry. Truth: "A wise man listens to advice." (Proverbs 12:15) — and sometimes the advice is to grieve, not grind.

Wisdom for Wounded Soldiers

If you're like me, you didn't grow up with a blueprint for a wise life. Maybe all you ever knew was survival. Maybe wisdom sounds like something for people who haven't seen what you've seen. Here's the truth Jesus doesn't wait for you to have it together before He teaches you to walk in wisdom. He meets you in your barracks, your rehab, your broken marriage, your sleepless nights. He shows you how to guard your heart, tame your tongue, and choose life over destruction. Wisdom is not about having all the answers, it's about knowing the One who does.

Chapter 25 Song Of Songs

Christ, the Bridegroom Who Redeems Intimacy

Chapter 25. Song of Songs

Christ, the Bridegroom Who Redeems Intimacy

"I am my beloved's and my beloved is mine." — Song of Songs 6:3

The Song of Songs is a love song unlike any other. Earthy and poetic. Passionate and pure. At first glance, it may seem like just a celebration of romantic desire, but beneath the surface is something much deeper, a picture of covenant love. A whisper of something eternal. This book pulls back the veil on a sacred mystery, the kind of intimacy we were created for not just with a person, but with God Himself.

That's the foundation of this chapter: before we can truly love another, we must first be loved by Christ. Before we can heal in a relationship, we must first be healed in Him.

Love That Isn't a Shortcut to Healing

Too many of us especially veterans carry wounds we don't even have words for. War strips us down. Trauma isolates us. Addiction numbs us. Shame silences us. We search for relief, comfort, and connection. Too often, we think a relationship will fix what's broken, but people aren't saviors. They were never meant to be.

I learned that the hard way I went from relationship to relationship looking for that thing that would heal my heart and soul. That is why I am so thankful for my wife Rayanne.

Before I came back into Rayanne's life her healing had already started. It started when Christ met her in her pain. Long before I reentered her life, she was already being made whole by the One who never left her side. Jesus became her safe place, her rescuer, her Bridegroom. She didn't wait to be loved by me, she had already been transformed by the love of Christ.

That's a crucial truth: healing doesn't come through another person. It comes through time in the presence of Jesus. Through surrender. Through intimacy with the only One who can truly restore what was broken. That is what Rayanne told me when we got back together after almost 20 years apart. She told me I needed to find healing first. She knew that we would never work if I just kept trying to put a person (her this time) in the place God needed to be. Boy was she right. Just another reason why I am so thankful for her.

The Garden Reopened

In the opening chapters of Genesis, man and woman walked naked and unashamed in the garden. Sin shattered that. Shame entered. Intimacy was fractured. In Song of Songs, we catch a glimpse of restoration: a garden reopened. A return to delight, devotion, and unashamed love.

"You are altogether beautiful, my darling; there is no flaw in you." — Song of Songs 4:7

That's not a statement about perfection, it's about perception redeemed by love. That's the kind of love Christ offers us. He sees the flaws, and still calls us beautiful. He knows the worst of us and still draws near.

Whether we're single, married, divorced, or widowed, our truest identity is not found in a spouse, it's found in our Savior. Jesus, the Bridegroom of the Broken

The Apostle Paul says:

"Husbands, love your wives, just as Christ loved the church and gave Himself up for her." — Ephesians 5:25

That's not just a model for marriage. That's a portrait of Jesus as the Bridegroom who pursues us relentlessly, delights in us even when we feel unwanted, speaks value over our scars, gives everything to make us His. He doesn't need us to be flawless. He doesn't flinch at our past. He simply says, "You are Mine."

For veterans who feel unlovable, distant, or disconnected from healthy intimacy this is the invitation, return to the One who already loves you perfectly. Not the bottle. Not porn. Not a person. Christ.

Our Story: From Ashes to Covenant

When Rayanne and I did reconnect, we weren't riding in on white horses. We were scared. We were humbled, and we were different. This wasn't a fairytale reunion, it was a testimony of grace.

We both knew what it was like to try to fill the void with all the wrong things. We both knew what it was to run from God and to come home limping, but now, because of Jesus. We had both tasted healing well I tasted it Rayanne was long past the appetizer and main course her healing was in full effect and I was just beginning my healing journey. The great thing is that when we stood together again, we didn't stand as two halves trying to make a whole. We stood as two individuals who had already been made whole in Christ, well one made whole (Rayanne) and one being made whole (me). Yes, God brought us back together but not to complete each other, but to reflect the completeness we already had in Him.

Now, our marriage is not the source of our healing, it's the fruit of it. We still have to choose grace. We still have to fight for intimacy that doesn't come naturally after trauma but there is joy again, laughter, trust, and a peace that didn't come from getting each other back, but from never losing Christ in the first place.

"Place me like a seal over your heart, like a seal on your arm; for love is as strong as death, its jealousy unyielding as the grave." — Song of Songs 8:6

You Are Already Loved

You don't need to find the right person to feel worthy. You don't need to fix your past to be loved, and you certainly don't need to be in a relationship to be whole.

You already have a Bridegroom and His name is Jesus. He calls you chosen. He calls you pursued, He calls you His own.

This is not just a truth for women or for church folks it's for warriors, for veterans, and for the weary. For the ones who've seen too much and felt too little for far too long. He is the intimacy your heart aches for, and His love is not a theory it's a covenant. You are His. And He is yours.

Chapter 26

Isaiah

Christ, the Suffering Servant and Sovereign King

Chapter 26. Isaiah

Christ, the Suffering Servant and Sovereign King

"But He was pierced for our transgressions, He was crushed for our iniquities; the punishment that brought us peace was on Him, and by His wounds we are healed." — Isaiah 53:5

Isaiah is often called "the fifth Gospel" and for good reason. Written over 700 years before Jesus walked the earth, this prophetic book describes His birth, life, death, and resurrection with such detail that it reads like an eyewitness account.

Isaiah doesn't just paint a portrait of the Messiah for scholars. He writes for the wounded, the oppressed, the addicted, the exiled, and the ashamed.

He writes for people like us, people who need more than religion, we need rescue. People who need more than a second chance, we need a new heart. People who don't just need answers, we need Jesus.

The God Who Sees Broken People

Isaiah's opening chapters read like a battlefield after the smoke clears. The people of God are scarred by rebellion, the land is sick with injustice, the leadership is corrupt, and the people are unclean:

"From the sole of your foot to the top of your head there is no soundness, only wounds and welts and open sores." — Isaiah 1:6

That's not just Israel's condition that was mine. Wounded by foster care trauma, covered in the scars of addiction, burned out by war, haunted by memories. Broken relationships, self-inflicted pain, and numbed emotions

Isaiah doesn't pretend these things don't matter. He doesn't look away. Neither does he leave us there. Because into the mess, Isaiah begins to sing of a Servant who would come… and carry it all.

Christ in Isaiah: The Suffering Servant

No other Old Testament book so clearly describes the crucifixion of Jesus. Isaiah 53 is the beating heart of the gospel:

"He was despised and rejected by men, a man of suffering, and familiar with pain." (v. 3) "Surely He took up our pain and bore our suffering…" (v. 4) "He was pierced for our transgressions…" (v. 5)

Jesus didn't come just to teach us He came to bleed for us. He didn't avoid suffering; He embraced it, so we could be healed.

Every veteran who's ever carried pain in silence needs Isaiah 53. Every addict who's relapsed and hated his or herself for it needs Isaiah 53. Every spouse who's tried and failed to love well needs Isaiah 53.

Because it reminds us: He already paid the cost. He already bore the shame. He already made peace for us. "By His wounds we are healed."

The King Who Still Reigns

Isaiah doesn't stop at the Cross he lifts our eyes to the Throne. In Isaiah 6, the prophet sees the Lord, "I saw the Lord, high and exalted, seated on a throne; and the train of His robe filled the temple." — Isaiah 6:1. Isaiah's response? "Woe to me! I am ruined." That's how it feels to stand in front of real holiness.

I've had moments like that sitting in the back row of a church, arms crossed, trying to play tough while the Holy Spirit broke through my walls. "Woe to me! I am unclean." That was the first honest thing I'd said in years.

I remember that day like it was yesterday.

At the time, I had just decided to go back to church. I was the president of a motorcycle club, wearing a tough, hardened mask so no one would see the broken man underneath. For three weeks straight, I parked outside that same church, sitting in my car, smoking weed, trying to work up the courage to walk through the doors.

On the third week, I finally went in. It just so happened to be baptism Sunday. Sixty-plus people gave their lives to Christ that day.

I sat in the back corner, out of the way, but the Holy Spirit came over me like a rushing wave. I started crying uncontrollably and fell to my knees in worship.

In that moment, I heard Him. "I'm all you need. I'm right here with you. I've always been here making a path, guiding each step, and watching your six." I did not really believe it yet but I wanted to. That was the day my faith journey truly began

Then came the fire. The cleansing. The commissioning.

"Your guilt is taken away and your sin atoned for… Whom shall I send?" — Isaiah 6:7–8

God doesn't just heal you, He calls you. He turns the addict into an ambassador, the foster kid into a leader, the soldier into a servant, the sinner into a son.

The Mission of the Messiah: Isaiah 61 and the Healing of the Brokenhearted

Isaiah 61 is like sunrise breaking over a war zone. After pages of judgment and lament, this chapter bursts in like a medic running into the smoke, kneeling beside the wounded, and speaking life:

"The Spirit of the Sovereign LORD is on me, because the LORD has anointed me to proclaim good news to the poor. He has sent me to bind up the brokenhearted, to proclaim freedom for the captives and release from darkness for the prisoners..."
— Isaiah 61:1

In Luke 4, Jesus reads this passage in the synagogue and says, "Today this Scripture is fulfilled in your hearing." This is His mission statement, and it's personal.

Binding Up the Brokenhearted

"Bind up" in Hebrew means to wrap and treat wounds with care. For many veterans, the deepest wounds never made it to the medevac. They went untreated because no one saw them. Isaiah 61 says Jesus sees every one of them, and tends to them Himself.

Proclaiming Freedom for Captives

Some chains are physical, but the most common are invisible addiction, shame, fear, and rage. Jesus doesn't just lighten the load He breaks the lock. You are not your failure, not your MOS, not your diagnosis, you are His.

Release from Darkness

Darkness can become familiar making you forget what light feels like. Jesus doesn't shame you for staying there He walks in and leads you out. This is restoration, not mere survival.

Beauty for Ashes

Ashes are what's left after the fire. Veterans know ashes burned out dreams, lost friends, and failed marriages, but in Christ, ashes aren't the end they're the soil for something beautiful.

The Oil of Joy for Mourning

Ancient oil marked honor and purpose. Jesus pours joy into those who thought they'd never smile again. The grief remains but hope rises with it.

From Broken to Builders

The chapter goes on to say, "They will rebuild the ancient ruins and restore the places long devastated." (Isaiah 61:4) That's our call take the places in your life that were wrecked and let God rebuild them so you can help others rebuild theirs.

The veteran who once hid from everyone becomes the mentor who walks a younger warrior through recovery. The addict who once burned every bridge becomes the one who helps others cross theirs. The one who almost gave up becomes the one who refuses to let anyone else fight alone.

When You Don't Feel Forgivable

I used to think I was too far gone too angry, too numb, too guilty but Isaiah showed me a Christ who says:

"Come now, let us settle the matter… Though your sins are like scarlet, they shall be as white as snow." — Isaiah 1:18

This isn't cheap grace it's blood-bought redemption. You don't have to carry the guilt anymore. You don't have to fix yourself first. He was crushed so you could be free.

Chapter 27

Jeremiah

Christ, the Weeping Prophet and the Promise of a New Heart

Chapter 27. Jeremiah

Christ the weeping prophet and the promise of a new heart

"Before I formed you in the womb I knew you, before you were born I set you apart." — Jeremiah 1:5

Jeremiah is known as "the weeping prophet." When you read his story, you understand why.

He didn't have an easy assignment. God called him young, before he had confidence, before he had approval. His mission? To preach to a rebellious nation, cry over their sin, and stay faithful when almost no one listened. He was misunderstood, isolated, imprisoned, and emotionally raw. Yet, he obeyed.

When the Call Feels Too Heavy

Jeremiah didn't ask for this. He didn't wake up thinking, "I want to be a prophet." But God had marked him before birth:

"Do not say, 'I am too young.' You must go to everyone I send you and say whatever I command you." — Jeremiah 1:7

That verse hits differently when you've lived through hell and still feel unqualified, when you've wrestled with addiction, disappointed people, and stared at your own broken reflection thinking, "God, You picked the wrong guy."

Like Jeremiah, I've learned: God doesn't call the clean, He cleans the called. He touches our mouths (Jeremiah 1:9), puts His words in us, and walks with us through rejection, depression, and spiritual battle.

Christ in Jeremiah: The Weeping Savior

Jeremiah's tears weren't weakness—they were prophetic. He wept over Jerusalem. So did Jesus: "As He approached Jerusalem and saw the city, He wept over it." — Luke 19:41

Jeremiah foreshadows the Savior who would feel what we feel:

Rejected by His own people

Accused by the religious

Betrayed, beaten, and buried

Yet still faithful

Jesus is the greater Jeremiah:

Not just a prophet with tears—but a Redeemer who takes on our grief.

Not just one who speaks God's Word—but the Word made flesh. When Jeremiah cries, "Is there no balm in Gilead?" (Jer. 8:22), Jesus answers: "I am the healing you've been waiting for."

For veterans, that balm is more than a metaphor, it's the slow stitching together of a torn soul, the steadying of hands that once shook with rage or withdrawal, and the softening of a heart hardened by loss.

Hope in the Dark: The Field in Anathoth

One of Jeremiah's most radical acts came while imprisoned: he bought a field (Jeremiah 32:6–15).

From a military standpoint, it made no sense. The Babylonians were besieging the land. Everything looked doomed. But God told him to do it as a sign of hope, that one day fields would again be bought, homes rebuilt, and life restored.

That image still stirs me. Sometimes faith means investing in a future you can't see yet. I've met vets who did this, enrolling in school while still battling PTS, starting a marriage even with scars, planting a garden when all they'd known was war. Faith is buying the field when the battlefield still smokes.

The New Covenant: A Heart Rewritten

One of the most powerful promises in Jeremiah comes in chapter 31: "'The days are coming,' declares the Lord, 'when I will make a new covenant… I will put My law in their minds and write it on their hearts. I will be their God, and they will be My people.'" — Jeremiah 31:31–33

That's Jesus. That's the Cross. That's what happened when He poured out His blood—not to reinforce rules, but to rewrite our nature.

This is why I'm no longer addicted—not by my strength, but because Jesus changed my desire. This is why I've stayed sober for almost three years without a single craving—not because I figured it out, but because grace reprogrammed my heart. This is why my wife and I are still together, healing, growing, and praying—because the covenant isn't written in ink, it's etched in mercy.

Bishop T.D. Jakes once said: "God is not intimidated by your tears. He counts them, bottles them, and waters your future with them." That's Jeremiah. That's the new heart. That's Jesus.

Chapter 28

Lamentations

Christ, The Hope That Rises from The Rubble

Chapter 28. Lamentations

Christ, the Hope that Rises from the Rubble

"Yet this I call to mind and therefore I have hope Because of the Lord's great love we are not consumed, for His compassions never fail. They are new every morning; great is Your faithfulness." — Lamentations 3:21–23

There's a unique kind of pain that doesn't just wound the body, it devastates the soul. Lamentations is written from the ruins of Jerusalem. Everything the people of God held dear was gone: the temple burned, homes destroyed, and families ripped apart.

The author who most believed it was Jeremiah, walks through the rubble and writes five poems of grief, of mourning, of holy lament. It's not a happy book, but it's an honest one, and sometimes, that's exactly what you need.

When Everything Breaks

There are seasons in life where everything collapses. I've lived through more than one.

When I was torn from my mother as a child and shuffled through 14 foster homes. When war left me with invisible wounds I didn't know how to name. When addiction stripped me of my identity, my peace, and almost my future.

Veterans know this feeling well. We've walked through cities torn apart overseas… and then returned home to find the cities of our own souls in ruins. The battlefield may be behind us, but the aftershocks keep shaking the foundation.

There's a kind of silence that sets in after those seasons. You sit in the ashes, unsure what to pray. You know God is real, but you can't feel Him. You're not even sure if you want to.

That's where Lamentations begins.

"He has walled me in so I cannot escape; He has weighed me down with chains." — Lamentations 3:7 "My soul is downcast within me." — Lamentations 3:20

Christ in the Book of Lamentations

Jesus isn't named in Lamentations, but He's present in every verse. When the city is crushed, Christ weeps over Jerusalem (Luke 19:41).

When the people cry out in abandonment, Christ echoes them from the Cross (Matthew 27:46).

When Jeremiah cries, "Is it nothing to you, all you who pass by?" (Lamentations 1:12) it's a prophetic preview of the Crucified Savior hanging exposed before a watching world.

When the poet sits in silence under the heavy hand of God (Lamentations 3:28) we remember Jesus' silence before Pilate (Mark 15:5), choosing submission to the Father's will over self-defense.

Lamentations teaches us that Jesus doesn't rush past pain.

He enters it. He sits in the ash heap. He weeps beside the ruined walls. He listens to the raw, wordless groans of a broken heart, and doesn't flinch.

But Then… Hope Breaks Through

In the middle of the darkest chapter of the book comes one of the brightest verses in all of Scripture: "Yet this I call to mind and therefore I have hope: Because of the Lord's great love we are not consumed, for His compassions never fail. They are new every morning; great is Your faithfulness." — Lamentations 3:21–23

That verse saved me more than once.

It reminded me that God's mercy doesn't expire. It's not based on my performance or consistency or strength. It's new every morning, even the mornings I didn't want to wake up.

When I was battling suicidal thoughts… When I relapsed and hated myself… When I sat in recovery groups not knowing what to say… When I failed my wife again… When I felt unlovable and unseen… Those mornings were met with mercy. Not judgment. Not condemnation. Mercy.

As John Eldredge writes in *Resilient*: "Hope is not a luxury; it's oxygen for the soul. Lose hope and you lose the very thing that keeps you going."

We live in a culture that rushes grief. Even the Church sometimes gets uncomfortable with pain that doesn't resolve in three Bible verses and a smile.

Lamentations teaches us something holy: It's okay to sit in the rubble. It's okay to cry out, "God, where are You?" It's okay to not understand. Jesus never rebuked the weeping. He joined them. He wept at Lazarus' tomb, knowing full well He would raise him.

So if you're in a season of silence… if your soul feels crushed… if the old wounds have reopened, you're not alone. Christ is there. He doesn't walk away. He waits, and when you're ready, He walks you out of the ruins. One step. One morning of mercy at a time. Your step today might just be whispering His name. That's enough for Him to meet you there.

Chapter 29

Ezekiel

Christ, The Breath in the Valley of Dry Bones

Chapter 29. Ezekiel

Christ, The Breath in the Valley of Dry Bones

"Then He said to me, 'Prophesy to these bones and say to them, dry bones, hear the word of the Lord!'" — Ezekiel 37:4

The book of Ezekiel opens with thunder and fire, strange wheels in the sky, four-faced creatures, and a vision of glory most people never talk about in Sunday school, but beneath the imagery, the message is clear: God is still speaking to the exiles.

This is a book written from captivity. Ezekiel is living among a people who had lost their home, their hope, and their sense of who they were.

I know that feeling. After 14 foster homes, I didn't know where I belonged. After the war, I didn't know who I was. After addiction, I didn't think God could use me. After relapses, I didn't think I deserved breath, but in Ezekiel, we see something powerful, Even in exile, God still shows up.

God Speaks in Strange Places

God comes to Ezekiel not in a temple, but by a river in Babylon. Not in a sanctuary, but in the shadows. Not in the land of the righteous, but in the land of the ruined.

That gives me hope. Because some of the most sacred encounters I've had didn't happen in a church. They happened in war zones. In recovery meetings and on the edge of relapse. In the bedroom, with my face in the carpet, whispering, "God, I can't do this anymore" and you know what? He met me there.

The Valley of Dry Bones

Ezekiel 37 is one of the most iconic chapters in the Bible and one of the most personal for me.

"The hand of the Lord was on me… He set me in the middle of a valley; it was full of bones." — Ezekiel 37:1 Not wounded bones. Not resting bones. Dead. Dry. Done.

That's how I felt at my lowest point spiritually dead, emotionally numb, hollowed out by trauma, sin, addiction, and shame. Then God asks a question: "Son of man, can these bones live?" — Ezekiel 37:3 The only honest answer? "Sovereign Lord, You alone know."

Then comes the command: "Prophesy to these bones… I will make breath enter you, and you will come to life." — Ezekiel 37:4–5

Christ: The Breath Giver

Jesus is the Word spoken over the bones. In the beginning, He breathed life into Adam (Genesis 2:7).

In the upper room, He breathed on His disciples and said, "Receive the Holy Spirit" (John 20:22).

On Pentecost, He sent His Spirit like a rushing wind to breathe life into the Church (Acts 2).

In recovery rooms, whispered prayers, and halfway houses He is still breathing. He didn't just raise me from addiction. He breathed life into the parts of me I thought would never feel again, my voice, my desire to live, my ability to love, my identity as a man of God, my marriage, and my future.

"So I prophesied as He commanded me… and breath entered them; they came to life and stood up on their feet a vast army." — Ezekiel 37:10

Veterans: Remember Who You Are

When I read "a vast army," I can't help but picture my brothers and sisters in uniform. Many of us left the service thinking our mission was over, especially if we left wounded, dishonorably discharged, or feeling like a failure.

God's army isn't made of perfect soldiers it's made of resurrected ones. Whether you've been home for decades or you're still in the fight for your life, you're still called, still chosen, still commissioned.

C.S. Lewis once wrote, "You are never too old to set another goal or to dream a new dream." In Christ, your orders aren't cancelled they're renewed. Your identity isn't "former soldier" it's "redeemed warrior."

From Bones to Breath

You may feel too far gone. You may think the damage is permanent. You may feel like the divorce, the addiction, the dishonorable discharge, or the criminal record has erased your future. Hear me: God speaks to bones, and not only does He speak He breathes.

The same Spirit who raised Christ from the dead is the Spirit who can raise you from addiction, numbness, despair, or disbelief.

You are not forgotten. You are not irredeemable. You are not too dry to rise. If He can rebuild Israel from the ashes, If He can call warriors from skeletons, He can resurrect you.

Chapter 30

Daniel

Christ, the Son of Man Who Stands in the Fire

Chapter 30. Daniel

Christ, the Son Of Man Who Stands in the Fire

"Look! I see four men walking around in the fire, unbound and unharmed, and the fourth looks like a son of the gods."

— Daniel 3:25

The book of Daniel is a story of resilience in exile.

Of boldness under pressure. Of faithful men standing their ground when the cost could be death. Daniel and his friends were taken from their homeland, given new names, stripped of their identity, and expected to bow, but they didn't. What gave them that kind of courage wasn't pride it was presence. The presence of the Son of Man, who walks in the fire.

Living in Babylon Without Bowing to It

Daniel lived in Babylon, but Babylon didn't live in him. That's hard to do. I've been in environments that tried to strip away who I was, Combat zones where survival came first. Recovery rooms where shame shouted louder than hope. Neighborhoods that didn't expect me to make it. Foster homes where faith was nowhere to be found. The pressure to conform, to blend in, to just "get through" was overwhelming. Daniel's story reminds us you can survive in Babylon without losing your identity in Christ.

Another in the Fire

Shadrach, Meshach, and Abednego stood before a furnace, flames so hot they killed the soldiers who came near. Their crime? Refusing to bow to a golden image. Their response still shakes the walls of fear today: "The God we serve is able to deliver us… But even if He does not… we will not serve your gods." — Daniel 3:17–18

They were thrown into the furnace bound, condemned, surrounded by flames, and then the unthinkable happened. The king leapt to his feet. His voice trembled "Didn't we throw three men in the fire? I see four… and the fourth looks like a son of the gods." (Daniel 3:25)

That fourth man wasn't an angel. It was the preincarnate Christ. The Son of Man. The One who promised, "I will never leave you nor forsake you."

I've been in fires that had nothing to do with flames. Nights when PTS flashbacks had me gasping for air. Days when cravings clawed at me until my hands shook. Courtrooms and hospital rooms where I felt completely exposed. Moments when my marriage hung by a thread. The heat was real. The danger was real. The shame was real, and so was He. I didn't walk through it alone. There was Another in the fire. That's the promise to every veteran, every addict, every spouse trying to hold it together, every believer hanging on by a thread: The fire will come, but you will not face it by yourself.

The Lions Can't Have You

Later in Daniel's life, he was thrown into a lions' den for refusing to stop praying. He didn't panic. He didn't negotiate.

He just knelt down, as he always had, and trusted God. "My God sent His angel, and He shut the mouths of the lions." — Daniel 6:22 The lions were real but they were restrained by God's hand. I've been surrounded by lions too, Accusations from the enemy. Voices telling me I'd fail again. Old temptations crouching at the door, but Jesus shut their mouths. Not because I was strong but because He is faithful.

Christ: The Son of Man

Daniel saw Him long before Bethlehem: "I looked, and there before me was one like a Son of Man, coming with the clouds of heaven." — Daniel 7:13

Jesus would later take that title as His own: "You will see the Son of Man coming on the clouds of heaven." — Matthew 26:64

The Son of Man is: The One who walks through fire. The One who shuts the mouths of lions. The One who rules every kingdom with mercy and truth. "God whispers to us in our pleasures, speaks in our conscience, but shouts in our pains: it is His megaphone to rouse a deaf world." — C.S. Lewis. The fire may be where you hear His voice the clearest.

Faith in the Fire

You may feel the flames right now. You may feel the lions' breath on your neck. You may not see the way out. But Daniel's story shows us you don't have to escape the fire to experience God's presence. You don't have to silence the lions yourself. You don't have to bow to Babylon. There's Another in the fire and if He's with you in the flames, the flames cannot consume you.

Chapter 31

Hosea

Christ, the Faithful Husband Who Never Stops Pursuing

Chapter 31. Hosea

Christ, the Faithful Husband Who Never Stops Pursuing

"The Lord said to me, 'Go, show your love to your wife again, though she is loved by another and is an adulteress. Love her as the Lord loves the Israelites…'" — Hosea 3:1

The story of Hosea is scandalous. God tells a prophet to marry a prostitute named Gomer knowing she will leave him, cheat on him, and sell herself to other lovers. When she does, God commands Hosea to go after her, to buy her back, to love her again.

Why? Because Hosea's love story is a living parable of God's relentless, undeserved love for His people. It's more than a parable. It's personal.

Unfaithful and Still Chosen

Hosea's marriage was a mess, just as Hosea went looking for Gomer in the back alleys and brothels, Jesus comes looking for us. Not when we had cleaned up, not when we were worth "redeeming" in human eyes, but when we were still running the other way. "I will heal their waywardness and love them freely, for My anger has turned away from them." — Hosea 14:4 This isn't weak grace. This is fierce, pursuing mercy.

Bought Back from the Auction Block

Hosea 3 paints a gut-wrenching picture: Gomer, degraded and enslaved, stands on the auction block perhaps half naked, eyes down, chains clinking with every small movement. Men bid not to cherish her, but to use her. Then, through the crowd, Hosea's voice cuts through the noise. He steps forward with fifteen shekels and some barley the price of a slave and says, "She's mine."

He doesn't shame her. He doesn't list her failures. He redeems her. "So I bought her…" (Hosea 3:2) "Then I told her, 'You are to live with me… and I with you.'" (Hosea 3:3)

That's exactly what Christ did for me. I had sold myself to addiction, to self-protection, to chasing approval. I was standing on my own kind of auction block thinking no one would bid for a mess like me, and then Jesus stepped forward. Not with silver or barley, but with His own blood.

For the Veteran, the Struggler, the Runaway

Hosea's story is for anyone who's been told they're "too far gone." For the veteran drowning in memories of war. For the addict who can't imagine sobriety. For the spouse convinced their marriage can't be restored. For the believer who feels too dirty to pray. Christ is not standing on the porch waiting for you to make it back. He's already walking into the darkness to find you.

Dietrich Bonhoeffer once wrote, "We must learn to regard people less in light of what they do or omit to do, and more in the light of what they suffer." That's how Jesus sees you, not by your failures, but by the wounds He came to heal.

You Can Always Come Home

You may feel like Gomer. You've wandered. You've relapsed. You've sold yourself for cheap comfort, but the Gospel is not about you finding your way back it's about Jesus finding His way to you. Paying the price. Calling you beloved.

If you're reading this with guilt heavy on your chest, hear this: The cost is already paid. Your place is already set. You can always come home.

Chapter 32

Joel

Christ, the Restorer of What Was Lost

Chapter 32. Joel

Christ, the Restorer of What Was Lost

"I will repay you for the years the locusts have eaten."

— Joel 2:25

The book of Joel begins with devastation. Waves of locusts sweep across the land like an invading army, devouring crops, stripping the fields bare, and leaving behind silence where there once was life, but it doesn't end there. In its short three chapters, Joel moves from ruin to redemption, from famine to fullness, from judgment to joy, and at the center of that transformation is a promise, a promise that finds its ultimate fulfillment in Jesus Christ.

He is the Restorer of what was lost.

For the veteran, the first responder, the addict in recovery, or the survivor of trauma, this is more than poetic language, it is a lifeline. When you've lived through years that feel wasted, when you've lost time to darkness, pain, or destructive choices, Joel's prophecy speaks directly to you God restores.

The Trauma That Devours

Trauma doesn't just touch the surface. It swallows things whole. It devours peace. It drains relationships. It disorients time. You blink and suddenly years have passed, years marked not by growth, but by survival. For many who served in combat, responded to disaster, or coped through addiction, these are the "locust years." They're the silent seasons no one celebrates years consumed by pain, regret, and internal war.

Joel gives voice to that loss. His words don't minimize it, they name it. In naming it, he also introduces hope. The devastation isn't the end. It becomes the soil where healing begins. Christ enters the ruins, not just to forgive, but to redeem. He is not merely the Savior of our souls, He is the Restorer of our years.

When you've lost time to drugs, to rage, to silence, to fear Christ says, "I will repay you." That's not just spiritual comfort. It's a personal promise from the One who has authority over time, trauma, and your story.

Christ, the God Who Restores

God's words through Joel are some of the most tender in all of Scripture: "I will repay you for the years the locusts have eaten." — Joel 2:25

He doesn't just say, "I forgive you." He says, "I will restore what you lost." That is the heart of Jesus. When He walked this earth, He didn't just cleanse lepers, He gave them back their dignity. He didn't just cast out demons, He gave people back their minds, their relationships, their futures. That's what He still does today.

For those in recovery, this hits close to home. When the fog begins to lift and we realize how much time has been lost, relationships strained, opportunities missed, and years numbed, it's easy to feel like the best is behind us. The great news is that in Christ, restoration begins now. He can grow fruit in barren soil. He can bring beauty from ashes. He can redeem what you thought was ruined beyond repair. He doesn't just give peace. He gives love. Joy. Patience. Kindness. Goodness. Faithfulness. Gentleness. And Self control. These are not rewards for the religious. They are the fruits of a Spirit filled life, even for those of us who thought we were too broken to bear anything good again.

The Spirit Who Fills the Empty Places

Joel's prophecy takes a surprising turn. After promising restoration, God doesn't just stop at repayment, He speaks of an outpouring.

"I will pour out my Spirit on all people… Your sons and daughters will prophesy, your old men will dream dreams, your young men will see visions." — Joel 2:28

This was fulfilled in Acts 2, when the Holy Spirit came upon the early Church, but it continues today. The Spirit is not reserved for the unblemished. He is poured out on the wounded, the weary, and the recovering. On those who never thought they'd be used by God again. For the combat vet with survivor's guilt. For the paramedic who can't forget what they saw. For the addict who thought they'd blown every chance. For the spouse, the friend, the parent who wonders if joy will ever return, He pours out His Spirit still. The same Spirit who hovered over the void in Genesis now hovers over the void in your life and speaks light into it.

Chapter 33
Amos

Christ, the Defender of the Oppressed

Chapter 33. Amos

Christ, the Defender of the Oppressed

"Let justice roll on like a river, righteousness like a never-failing stream!"— Amos 5:24

The prophet Amos didn't come from power or pedigree. He wasn't a priest or a prophet's son he was a shepherd. A blue-collar man working the land, and yet God chose him to speak into corruption, injustice, and religious hypocrisy. His message was bold. His tone was fierce. His heart was on fire for the poor, the abused, the forgotten, and the exploited.

In Amos, we catch a glimpse of Christ, the Righteous Judge and the Merciful Defender. He is not indifferent to suffering. He is not silent in the face of injustice. He stands with the broken and calls out what's wrong. He defends the weak and restores what's been misused, and He calls us especially those who have worn the uniform to do the same.

The Warrior Who Speaks for the Voiceless

Amos stood on the front lines of spiritual war, not with weapons, but with truth. He confronted kings and religious leaders who wore the name of God but lived with no regard for the wounded among them. He saw how the powerful used the system to their advantage while the poor were crushed beneath it.

Veterans, first responders, and those in recovery know something about this kind of courage. Whether on the battlefield, in burning buildings, on overdose calls, or in the wreckage of trauma, we've stood in places where others ran. We've tried to protect the vulnerable, defend the innocent, and uphold justice, even when it cost us.

Amos reminds us true justice doesn't flow from human effort, it flows from the heart of God, and Christ is that heart made visible. In Him, justice and mercy meet. In Him, the oppressed are lifted up and the self-righteous are brought low. In Him, the bloodied and broken find someone who doesn't just defend them but delivers them.

If you've ever felt unseen while trying to hold the line… if you've ever served and bled for a world that still seems corrupt… if you've ever questioned where justice went look to Christ. He hasn't forgotten the cries of the wounded. He hears yours too.

Christ, the Justice Bearer and Peacemaker

In Amos 5, the prophet doesn't pull punches. He rebukes the nation for singing songs to God while turning a blind eye to suffering. He tells them their worship is meaningless without justice, but he doesn't leave them hopeless. He gives a vision of what God desires:

"Let justice roll on like a river, righteousness like a never failing stream!" This isn't about performance, it's about presence. Justice rolls because Christ walks with the hurting. Righteousness flows because He cleanses what we cannot. In Jesus, we see justice, not as cold judgment, but as compassionate restoration.

For the veteran drowning in guilt, for the medic who saw too much, for the addict who hurt others while trying to survive, Jesus is not standing with a gavel. He is standing with open arms. He defends not because we're innocent, but because He is merciful, and He took the punishment we deserved so we could receive the peace we never earned.

In Him, righteousness becomes more than a standard it becomes a stream that washes away shame and makes all things new.

Calling the Redeemed to Be Restorers

Amos isn't just about pointing fingers. It's a call to live differently. Those who have been rescued are called to become rescuers. Those who have received mercy are called to extend it.

Many of us have stood in dark places. We've been surrounded by pain, anger, injustice, addiction, and grief. Now that Christ has brought us into the light, we carry that light into the same dark places. We become agents of justice not through anger, but through compassion. Not with force, but with faith.

We listen to the cries of the oppressed because Christ listened to ours. We stand with the hurting because Christ stood with us. We speak for the voiceless because Christ spoke our name when no one else would.

Chapter 34

Obadiah

Christ, the Judge Who Humbles Pride and Heals Division

Chapter 34. Obadiah

Christ, the Judge Who Humbles Pride and Heals Division

"The day of the Lord is near for all nations. As you have done, it will be done to you…"

— Obadiah 1:15

Obadiah is the shortest book in the Old Testament, just one chapter. But don't let its size fool you. This book punches hard. It speaks directly into the ruins of rivalry, betrayal, and prideful injustice. The prophet delivers a sobering word of judgment, particularly against Edom, the nation that stood by and gloated when their relatives in Judah were attacked.

Even in a book about judgment, we find Jesus. In Obadiah, we see Christ as the righteous Judge who humbles pride, exposes betrayal, and brings healing to generational wounds. For veterans, first responders, and anyone who has experienced betrayal, broken trust, or deep division, this message hits close to home.

Obadiah is a reminder that justice is not forgotten, pain is not dismissed, and pride will not have the final word. Christ sees. Christ judges rightly. Christ restores.

The Pride That Destroys—and the Peace That Heals

Edom and Israel shared a long history. Their roots went all the way back to Jacob and Esau, two brothers torn by pride and rivalry. That fracture carried through generations. So, when Jerusalem fell, Edom didn't offer help. They cheered from the sidelines. They looted what remained. They reveled in their brother's ruin.

Obadiah's words confront that smugness head on. Pride, he says, is deceptive. It convinces us we're safe while our hearts grow cold. It makes us feel strong while our souls rot inside. It tells us justice is ours to dish out, when only God sees the full truth.

If you've ever watched betrayal unfold in your unit, your department, your family, or even your own heart, you've seen what pride can do. It destroys. It divides. It defends self over others, and yet… Christ enters even that brokenness.

He's not only the one who humbles the pride. He's the one who brings peace where division ruled. He tears down walls between enemies. He reconciles what generations of pain could not. He restores trust, first in God, and then slowly, painfully, with others.

In Him, even brothers divided by war like Jacob and Esau can be reconciled again.

Christ, the Righteous Judge and Merciful King

Obadiah declares that "the day of the Lord is near." That phrase can sound terrifying, but for the wounded and oppressed, it is a promise of justice. Not revenge. Not human rage, but holy justice the kind that puts things right.

For those who've suffered betrayal or carried deep injustice, it's good news to know that God does not ignore what's been done. Christ is not indifferent to what you endured. He knows the full story. He sees every wound and every act of prideful destruction, both the ones done to you and the ones done by you, but Christ doesn't stop at justice. He offers mercy to all who turn to Him. He gives grace to the broken and peace to the divided. He tears down the high places of pride not to destroy us, but to rescue us from self-destruction.

For those in recovery especially those wrestling with regret over how pride shaped your decisions or relationships, Christ speaks a better word. His humility heals. His truth disarms. His grace rebuilds.

From that place of healing, He empowers us to live differently to forgive what once seemed unforgivable and to walk humbly, no longer needing to win or be right, but to be redeemed.

Chapter 35
Jonah

Christ, the Greater Prophet Who Brings Mercy

Chapter 35. Jonah

Christ, the Greater Prophet Who Brings Mercy

"But the Lord provided a great fish to swallow Jonah, and Jonah was inside the fish three days and three nights."

— Jonah 1:17

I've known what it feels like to run from God. Not just wander or drift, but flat out run. To know deep in my bones what He was calling me to do, surrender, come home, forgive, speak up, and to choose the opposite. I've sat in rooms high out of my mind, knowing better. I've poured shots with my fist clenched in anger, praying half-hearted prayers I didn't mean. I've been in places where I should have died, not because the world was against me, but because I was against myself, and just like Jonah, God never stopped pursuing me.

The book of Jonah is about more than a fish. It's about a Father who chases us into the storm, into the darkness, even into the grave, to bring us back. Jonah ran. I ran. Maybe you're running too. But Christ, the Greater Jonah doesn't abandon us in the chase. He enters the storm with us, and He brings us out alive.

Running from the Call, Caught by Grace

Jonah was a prophet, but he didn't want to speak to Nineveh. They weren't just pagans they were enemies. The kind of people who had oppressed, murdered, and mocked Jonah's people, and God said, "Go." Jonah said, "No." So he ran.

I get it. There were people I didn't want to forgive. Places I didn't want to return to, especially in my mind. I thought I was justified in my anger. I thought some people didn't deserve mercy, especially me.

You can't outrun a God who still loves you.

When Jonah was swallowed by the fish, it wasn't punishment, it was protection. It was God giving him space to pray, to break, and to wake up. I've had my own belly of the fish moments, jail cells, detox, cold basements, silent drives home after everything fell apart. That's where I learned, God doesn't wait for us to clean up before He reaches down. He meets us in the deep.

"In my distress I called to the Lord, and He answered me." — Jonah 2:2

Even in rebellion, He answers. Even in relapse, He rescues.

As Philip Yancey wrote:

"Grace, like water, flows to the lowest part."

When Mercy Feels Offensive

Eventually, Jonah obeyed. He went to Nineveh and delivered the message:

"Forty more days and Nineveh will be overthrown." — Jonah 3:4

It wasn't eloquent. It wasn't even kind , but something miraculous happened, they repented. The king tore his robes. The people fasted. They turned, and God forgave and Jonah? He was furious.

That part hit me hard in recovery. I wanted grace for myself, but not always for others. I wanted justice, but Christ the greater Jonah whispers a better word. He doesn't just offer mercy. He teaches us how to extend it.

"To be a Christian means to forgive the inexcusable because God has forgiven the inexcusable in you." — C.S. Lewis

Sitting Under the Vine, Holding onto Bitterness

After Nineveh repented, Jonah sat outside the city, sulking. God provided a vine to shade him. Then God took it away, and Jonah lost it. "Is it right for you to be angry?" — Jonah 4:4

I've asked myself that question a hundred times. When bitterness has grown like a weed in my soul, when I've used old wounds as an excuse to stay angry, or distant… God's question still echoes. Is it right for me to be angry when He's been so kind?

I've seen war. I've buried friends. I've hated people. I've watched the worst of humanity unfold in real time, but the cross of Christ still stands. The mercy that saved me doesn't stop with me. If I want real peace not just survival, but shalom, I have to let go of the vine.

"Resentment is like drinking poison and then hoping it will kill your enemies." — Anne Lamott

Bitterness doesn't protect us. It poisons us. Mercy is the antidote.

Grace That Finds You in the Dark

Have you run from God? Are you sitting in bitterness, angry about what someone else got away with? Do you think your story is too far gone to be useful?

Jonah tells the truth we all need to hear: You can't outrun mercy. Even in your deepest regret, Christ is there. Even in your anger, He offers peace. He isn't afraid of your mess. He's already entered it and come out the other side. He's not asking you to pretend. He's asking you to follow, not in pride, not in performance, but in grace.

Chapter 36

Micah

Christ, the Humble King Who Walks with the Wounded

Chapter 36. Micah

Christ, the Humble King Who Walks with the Wounded

"He has shown you, O man, what is good. And what does the Lord require of you? To act justly, to love mercy, and to walk humbly with your God."

— Micah 6:8

I didn't grow up thinking I'd end up where I did, addicted, angry, haunted by war, wrecked by regret. I thought I was doing the right things. I thought I was strong. I served. I fought. I survived. But somewhere along the way, survival turned into bitterness, and bitterness turned into rebellion.

By the time I hit bottom, I wasn't just running from God I was doing everything I could to silence His voice. Spoiler alert it never worked.

Micah's voice is a lot like the one I tried to ignore. He doesn't speak with polished religious language. He speaks like a prophet standing in the ruins, calling out corruption, pleading for justice, grieving the silence of God's people in the face of suffering, and in Micah, I met the Christ I didn't expect, the humble King who doesn't just demand justice but walks with the broken to bring it.

The Shepherd Who Sees the Mess

Micah saw a society that had forgotten God. The wealthy trampled the poor. The courts were corrupt. The religious leaders cared more about profit than people. Sound familiar? It did to me. It looked a lot like the broken systems I had seen during and after the war. It looked like the darkness I saw in the eyes of some leaders who never cared about the soldiers under them. It looked like the system I turned to when I numbed out on drugs and alcohol, hoping no one would see how badly I was falling apart inside.

Micah doesn't just speak judgment he sees hope. He speaks of a Shepherd who would rise from a forgotten town called Bethlehem. Not a king with a crown and a throne, but a King with scars and a cross.

"But you, Bethlehem… out of you will come for Me one who will be ruler over Israel… He will stand and shepherd His flock in the strength of the Lord." — Micah 5:2, 4

That Shepherd is Jesus and He didn't run from the mess, He entered it. He stood with the oppressed. He walked among the outcasts. He touched the unclean, and He still walks with us, with the veteran who wakes up choking on memories, with the addict who's trying to stay clean, with the broken who still doesn't know how to pray.

A Veteran's Struggle With Justice and Mercy

Micah 6:8 is one of the most well known verses in the Bible. I heard it before. But I didn't really hear it until everything in my life had been stripped down. I remember reading it one night still trying to convince myself that God wasn't done with me.

"Act justly. Love mercy. Walk humbly with your God."

I read it again and again, and I broke. Because I had done the opposite. I had acted out of pride, not justice. I had loved revenge, not mercy. I had walked with shame, not humility. I had tried to carry my recovery like a badge instead of surrendering it like a broken offering, and still, Christ didn't walk away. He just kept walking beside me.

Veterans and first responders carry a unique weight. We've seen injustice. We've felt the pressure to act strong even when we're dying inside. Sometimes justice feels like punishment. Sometimes mercy feels impossible. Sometimes walking with God feels like crawling through sand, but Micah calls us back to the simplicity of it. You don't have to perform. You just have to walk, and you're not walking alone.

When Trauma Tries to Take the Fruit

I remember one night not long after I came home, sitting in my mustang, hands shaking, heart racing, parked in a church parking lot. I didn't want to go in. I didn't want to talk. I didn't want to be vulnerable. I just wanted peace, but peace doesn't come by avoiding the pain. It comes when we walk through it with Christ.

Micah shows us the fruit of walking with Him not just survival, but healing. Not just forgiveness, but transformation. That means love where there used to be rage, joy where there used to be numbness, peace where there used to be panic, patience where there used to be flashbacks.

Chapter 37

Nahum

Christ, the Avenger of Justice and Peace

Chapter 37. Nahum

Christ, the Avenger of Justice and Peace

"The Lord is good, a refuge in times of trouble. He cares for those who trust in Him." — Nahum 1:7

There was a time in my life when I didn't want mercy, I wanted revenge. I didn't want peace, I wanted payback. I had watched innocent people suffer. I had seen corruption win. I had felt abandoned by systems, by people, and maybe even by God. My soul was worn thin from war, betrayal, addiction, and loss, and deep down, I wondered: Will justice ever come?

Then I met Christ in the pages of Nahum.

Most people skip this little book. It's fierce. It's bloody. It reads like a war song, but for someone who's been through real battles, who's carried real wounds, Nahum offers something rare and powerful: the promise that God does not ignore evil, and He will not let injustice stand forever. Christ is not just the Lamb, He is the Lion who defends what is His.

Justice for the Wounded

Nahum prophesied the fall of Nineveh, the same city God had spared in Jonah's day after they repented. This time, Nineveh had returned to its violence. They had crushed nations. They had spilled innocent blood, and God said, Enough.

For the veteran who watched their brothers fall, for the first responder who couldn't save the child, for the trauma survivor who still hears the lies of their abuser, Nahum speaks a truth we long for: God does not forget.

"The Lord takes vengeance on His foes and vents His wrath against His enemies." — Nahum 1:2

It may feel like evil gets the final say. It may seem like the wicked prosper and the wounded are left bleeding. But Nahum shows us a Christ who is not passive. He doesn't look away. He doesn't shrug off injustice. He sees. He remembers and He avenges.

As C.S. Lewis wrote in *The Problem of Pain*:

"God is not proud. He will have us even though we have shown that we prefer everything else to Him… But He is no tame lion."

We worship the Lamb, yes but also the Lion of Judah.

When You've Been the Enemy, Too

Here's the twist, we've all been Nineveh.

I've hurt people. I've said things in anger that wounded people I loved. I numbed out and disappeared when people needed me present. I've lied. I've manipulated. I've used. I may have seen evil done to me, but I've also done wrong myself.

That's why Christ came. Not just to bring vengeance, but to bear it. On the cross, He took the full justice of God, not to excuse sin, but to defeat it.

"Mercy triumphs over judgment." — James 2:13

In Christ, the judgment we deserved fell on Him, and now, He offers us more than justice, He offers us peace. Not just peace with God, but peace within ourselves, peace with our past, peace with those who hurt us.

That's the real miracle. That the God who thunders against injustice can whisper peace into the soul of a man who once longed for violence.

Refuge in the Middle of the Storm

Nahum doesn't only show the wrath of God. It shows His heart:

"The Lord is good, a refuge in times of trouble. He cares for those who trust in Him." — Nahum 1:7

I've needed that refuge. Maybe you do too. When the guilt still stings, when the flashbacks still come, when anger feels easier than healing, Christ is a shelter, not a storm. He doesn't blow you down, He covers you. He doesn't shame you, He defends you.

That's what makes Him trustworthy, not just that He judges rightly, but that He suffers with us until peace returns.

Chapter 38

Zephaniah

Christ, the Mighty One Who Saves in the Fire

Chapter 38. Zephaniah

Christ, the Mighty One Who Saves in the Fire

"The Lord your God is with you, the Mighty Warrior who saves. He will take great delight in you; in His love He will no longer rebuke you, but will rejoice over you with singing."

— Zephaniah 3:17

The book of Zephaniah begins with fire and ends with a song. It opens with judgment, God's holy anger against corruption, idolatry, and injustice, but woven through the warnings is a profound promise: the Lord is not just a righteous Judge; He is a mighty Savior. His justice clears the way for restoration. His fire is not just for destruction, it's for refining.

As someone who's walked through the fire of war, addiction, and shame, I used to see God's judgment as a threat. Now, I see it as part of His mercy, because He burns away the lies we've believed about ourselves. He confronts the idols we've trusted, and in the ashes, He builds something new.

From Collapse to Cleansing

Zephaniah's prophecies strike hard. The nation had turned from God. The leaders were corrupt. The people had grown numb. The vulnerable were exploited. In a sense, they were in moral and spiritual collapse, a situation not unfamiliar to many veterans, first responders, or those in addiction recovery.

When you've lived through trauma, chaos becomes normal. Silence feels dangerous. Hypervigilance replaces rest. That's not just a mental state, it becomes a spiritual one too. We forget what peace even feels like. The fire Zephaniah speaks of is meant to jolt us awake. Here's the beauty: God doesn't stop at warning. He promises to purify. He promises to restore. He promises to dwell with His people again. It's not just a fire of judgment, it's a fire of love that refuses to leave us in ruin.

The God Who Sings Over the Broken

Zephaniah 3:17 is one of the most tender, powerful verses in the entire Bible. Picture this: the same God who thunders in judgment also sings over you with joy. This is for every soldier who thinks they're too far gone. This is for every medic who still hears the screams at night. This is for every addict, every survivor, every fighter still holding on. God doesn't just tolerate you, He delights in you. He draws near, not because you're strong, but because you're His.

This is how healing begins not by numbing the pain, but by hearing the song. A real song. A rescue song from the One who enters the fire with you.

Hope for the Battle Weary

Zephaniah calls out a proud, self-sufficient people, but for the humble, for the brokenhearted, there's an invitation to shelter. In Christ, the Judge becomes our Advocate. The Warrior becomes our Shepherd. The Lion becomes the Lamb.

In Him, we begin to taste the fruit of the Spirit again, peace where there was chaos, joy where there was despair, gentleness in place of rage, faithfulness that carries us when we can't carry ourselves.

Chapter 39

Haggai

Christ, the Glory Who Fills the Empty Temple

Chapter 39. Haggai

Christ, the Glory Who Fills the Empty Temple

"The glory of this present house will be greater than the glory of the former house," says the Lord Almighty.

"And in this place I will grant peace."

— Haggai 2:9

The book of Haggai is a call to rebuild. After years in exile, the people of Israel returned home to find their temple in ruins. At first, they began to rebuild, but discouragement, distractions, and fear crept in, So, they stopped. They built their own houses but left God's house in shambles.

Sound familiar?

As veterans, first responders, or people in recovery, we often return from battles with our souls in ruins. We try to move forward, to rebuild our lives, but something always gets in the way guilt, addiction, anxiety, chronic pain, and spiritual exhaustion. We start off with fire, but when it gets hard, we put the bricks down and go silent, but through the prophet Haggai, God speaks to the discouraged:

"Be strong… I am with you." (Haggai 2:4)

He doesn't just command us to rebuild, He promises to dwell in what we rebuild, fill it with glory, and give us peace.

The Pain of Incomplete Healing

In Haggai's day, the people were haunted by comparison. The old temple had been magnificent. The new one felt small and pathetic. Some wept as they laid its foundation because it reminded them of what had been lost.

If you've ever tried to rebuild your life after trauma, you know this feeling. Maybe you look at the life you once had or the life you wish you had and it feels like whatever you're trying to build now doesn't measure up. There's shame in your recovery. Guilt in your rebuilding. You're haunted by what could have been. Here's God's promise: the glory of the present house will be greater.

Christ in the Rubble

Haggai is ultimately about a temple, but the true Temple was never bricks and gold. The true Temple is Jesus. John 1:14 says, "The Word became flesh and dwelt among us," literally meaning He "tabernacled" or "pitched His tent" in our midst.

Christ is God with us in the ruins, and now, through His Spirit, we are called His temple. (1 Corinthians 6:19) He lives in us. Not because we're clean, but because He cleanses. Not because we're worthy, but because He is faithful.

Haggai reminds us that God sees what we start, and He calls us to finish, not in our strength, but in His. When we rebuild with Christ, it's not just recovery, it's resurrection. Start again. He's not waiting to punish. He's waiting to fill your new foundation with His glory.

Chapter 40

Zechariah

Christ, the Pierced King Who Brings Peace

Chapter 40. Zechariah

Christ, the Pierced King Who Brings Peace

"They will look on Me, the one they have pierced, and they will mourn for Him as one mourns for an only child…"

— Zechariah 12:10

Zechariah is a book of visions. It's filled with horses and horns, lamps and flying scrolls, crowns and fountains. At first glance, it might feel overwhelming, but at its heart, it's about hope for the broken and the promise of a coming King who will heal, restore, and reign in peace.

Zechariah speaks to a people who had seen war, exile, and spiritual fatigue. They were rebuilding a ruined city and trying to trust that God hadn't abandoned them. Sound familiar?

Zechariah says: The King is coming, and His hands bear the proof that He understands pain.

The Pierced One Who Understands Our Wounds

Long before the crucifixion, Zechariah saw it:

"They will look on Me, the one they have pierced…" (Zechariah 12:10)

This is one of the clearest prophecies of Jesus in the Old Testament. The Messiah would come not as a distant ruler but as one who suffers with and for His people. He would be pierced. He would carry sorrow. He would know grief.

That means Christ understands your trauma. He knows what it's like to be misunderstood, abandoned, wounded, and rejected, and because He was pierced, He alone can bring peace to pierced souls. He doesn't just sympathize, He redeems. He doesn't just patch you up, He makes you new. Whether your wounds are visible or hidden, fresh or decades old, His wounds speak louder. Through those wounds, peace flows like living water.

A Warrior King Who Brings a Gentle Reign

Zechariah sees Christ not only as pierced but as a coming King, riding in humility:

"See, your king comes to you, righteous and victorious, lowly and riding on a donkey…" (Zechariah 9:9)

This prophecy was fulfilled when Jesus entered Jerusalem, Palm Sunday. The crowd expected a military conqueror, but He came as a suffering servant. His victory wouldn't come through war, but through surrender. Through a crown of thorns, not a golden one, and through the cross.

To those of us trained to fight, to take the hill, to always stand guard, this seems upside down, but Christ's reign is upside down: the last are first, the broken are blessed, and peace comes through surrender, not domination. His gentleness doesn't make Him weak. It makes Him safe. Safe for wounded souls. Safe for warriors who can't rest. Safe for those who've been on edge so long, they've forgotten what peace even feels like.

A Fountain for the Unclean

Zechariah gives another vivid image of Christ's mercy:

"On that day a fountain will be opened… to cleanse them from sin and impurity." (Zechariah 13:1)

You don't have to wash yourself before coming to Him. He is the Fountain. He is the cleansing. He is the welcome, and His arms are open. The blood of Jesus is not just enough to cover your sins, it's enough to soak through every part of your past, your habits, your history, your hurt. Whether your trauma is combat, childhood, divorce, addiction, betrayal, or all the above, this Fountain is open, and it never runs dry.

Hope for the Battle-Worn

Zechariah's message isn't just prophetic, it's pastoral. It speaks deeply to those who have been through hell and wonder if God still has a purpose for them. The people Zechariah addressed were trying to rebuild a life from the rubble. The temple was in ruins. Their identity as God's people felt shaky. They had seen loss, exile, and failure. But Zechariah speaks into that discouragement with vision after vision that says, God is not done with you yet.

That same voice echoes to the veteran haunted by war, the survivor carrying years of buried trauma, the addict trying to stay clean one more day. Christ does not discard the wounded, He redeems them. He doesn't overlook the tired builder still laying bricks, He crowns them with purpose.

Zechariah 4:6 delivers a battle cry that doesn't rely on our own strength: "Not by might nor by power, but by My Spirit,' says the Lord Almighty."

You may not feel strong enough to rebuild. You may look at your past and see too much damage, but God sees a temple rising. He sees beauty where you see rubble. Your healing will not come by white-knuckled effort or pretending everything is okay. It will come by His Spirit, slowly, deeply, truly.

Chapter 41. Malachi

Christ, the Sun of Righteousness Who Rises with Healing

"But for you who revere My name, the sun of righteousness will rise with healing in its rays…"

— Malachi 4:2

The final book of the Old Testament ends not with a neat bow, but with a burning need. Malachi is a prophet calling people out, priests who had grown lazy, hearts that had grown cold, and communities that had lost their fear of God. His words sting with conviction, but they also shine with promise. He points to the coming day of the Lord and to the One who would rise like the sun.

For the weary, the wounded, the ones in recovery, and the ones still in the trenches of trauma Malachi reminds us that the dark doesn't get the last word. The dawn is coming, and His name is Jesus.

Wounds That Wait for Healing

Many veterans and first responders understand the ache of waiting. Waiting for help. Waiting for peace. Waiting for something to finally work. We go to meetings. We show up for therapy. We pray. We fight, and still some mornings feel darker than others.

Malachi was written to people who had waited too. God's promises felt delayed. His presence felt distant. Their worship had grown routine, and their hope had worn thin, but even in that fog, Malachi speaks a word to those who fear the Lord:

"The sun of righteousness will rise with healing in its wings." (Malachi 4:2)

That's not just poetic, it's prophetic. Christ is that sun, and His light doesn't just expose, it heals. He comes not just to forgive but to restore, to revive, to renew. The sunrise of Christ cuts through the night of addiction, depression, bitterness, guilt, and shame, and in His light, we see clearly again: we're not too far gone.

The Fire and the Furnace

Malachi also speaks of fire:

"Who can endure the day of His coming? He will be like a refiner's fire…" (Malachi 3:2)

Some of us have lived through literal fire, combat, burning buildings, explosions. Others have walked through emotional fire, divorce, loss, betrayal, relapse., and sometimes, when we meet God, we expect Him to extinguish every flame instantly.

Malachi says: this fire purifies. Christ doesn't always remove the heat. Sometimes He uses it to refine us. He burns away the lies, the false identities, the toxic coping, and the pride. It hurts, but it heals.

As A.W. Tozer wrote:

"It is doubtful whether God can bless a man greatly until He has hurt him deeply."

Pain isn't the end of the story. Purity is. The Hope Beyond the Silence After Malachi, there's silence, 400 years of it. No prophets. No fresh revelations. Just waiting.

Many of us know that kind of silence. We've cried out and felt nothing. We've knocked and heard no answer. We've begged for healing, and the heavens seemed still. Silence is not absence. Delay is not denial. The dawn always takes a little time after the darkest hour. Malachi is proof that even when we don't feel it, God is preparing something. The Sun of Righteousness is rising, and when He does, healing will ride on His rays.

The Messenger and the Mission

Malachi doesn't just speak of the Sun of Righteousness, he also prophesies a forerunner, a messenger who will prepare the way for the Lord's arrival. That messenger, as we learn later, is John the Baptist.

"I will send My messenger, who will prepare the way before Me..."
— Malachi 3:1

Why does that matter? Because before healing comes, God often sends a voice, a call to turn, to prepare, and to surrender. For veterans, first responders, and anyone recovering from trauma or addiction, that voice may come through a sponsor, a counselor, a pastor, or even a stranger who says something that pierces the fog. It may not be dramatic, but it is divine. It's God breaking the silence.

God sends messengers because He loves us too much to leave us in numbness. He stirs our hearts. He disrupts our cycles. He reminds us that we're not alone and that He is on the move.

Here's the twist: you may be the messenger, too.

Malachi isn't just about personal healing, it's about preparing others. When you've been through the fire and found the Fountain, when the dawn has broken over your dark night, you become a lighthouse for others. You get to point to the coming

King. You get to say, "He met me in the silence. He lit up my darkness, and He will do the same for you."

So don't despise your waiting. Don't waste your wounds. The Sun is rising, and He's not just healing you, He's commissioning you.

Chapter 42
The Silence Before the Savior

Holding on to Hope in the Waiting

Chapter 42. The Silence Before the Savior

Holding on to Hope in the Waiting

"Though it linger, wait for it; it will certainly come and will not delay."

— Habakkuk 2:3

Four hundred years. That's how long the people of God waited between the closing of Malachi and the opening of Matthew. Four centuries without a prophet. Four centuries without a fresh word from heaven. Four centuries of silence.

It's hard to describe how heavy that silence must have felt. Generations were born and died without hearing the voice of God. Empires rose and fell. Scribes clung to old scrolls. Families lit Sabbath candles and told stories of deliverance, and all the while, the heavens remained still.

Silence doesn't mean absence, and delay doesn't mean denial. God may have been quiet, but He was never gone. He was setting the stage. He was preparing the way. He was weaving together a story far greater than they could see.

When Heaven Feels Silent

Many of us know what that silence feels like. Veterans waiting on disability paperwork. First responders trying to numb the weight of what they've seen. Addicts counting days and relapses. Trauma survivors asking: Where is God in all of this? You go to meetings. You sit in church. You cry out in prayer. But it feels like no one's home. It's in those seasons when the enemy whispers, "God has left you." Scripture shouts a better word:

"I will never leave you nor forsake you." (Hebrews 13:5)

God doesn't ghost His people (though he does send His Holy Ghost. Sorry seemed like a good dad joke). He doesn't give up on the wounded. He draws near, especially in the silence.

The Stillness Before the Breakthrough

Those 400 years weren't wasted. They were filled with purpose. Rome built roads that would one day carry the gospel. Languages shifted so that the New Testament could reach the world. A young virgin was born and would soon whisper yes to the angel's call, and then, light broke into the darkness.

Jesus came not into a world overflowing with faith, but into one suffocating under religion and Roman rule. He didn't wait for perfection, He came into the pain. That's how He works. He comes to us in the silence. In the pain. In the places we feel forgotten.

The Waiting Is Sacred

Maybe you're in your own 400 years of silence. Maybe your faith feels like a dry desert. Maybe your prayers feel like they hit the ceiling. Maybe your recovery feels stalled, or your marriage feels distant, or your body is worn out from service, surgery, or self-destructions.

Listen, brother. Sister. The waiting is not wasted. God does His best work in the dark. The seed grows before it sprouts. The stars shine before the sunrise. The tomb sealed for three days was the doorway to resurrection. As C.S. Lewis once said: "God whispers to us in our pleasures, speaks in our conscience, but shouts in our pains: it is His megaphone to rouse a deaf world." Hope Never Stops Hope isn't the absence of pain. It's the choice to believe that pain isn't the end of the story. Even in the silence, Christ is present. Even in the waiting, the Holy Spirit groans with us (Romans 8:26). Even in the dark, the dawn is on the way.

For the veteran with too many nightmares. For the firefighter who can't unsee the flames. For the nurse who hides her tears. For the addict who keeps falling. For the one whose family has given up. God hasn't. Christ came once through 400 years of silence. And He'll come to you again, even now.

Resources for the Wounded, Weary, and Waiting

You don't have to wait alone. Here are places to reach out for help today:

Veterans & First Responders

Veterans Crisis Line — Dial 988, then press 1 or text 838255

https://www.veteranscrisisline.net

Mighty Oaks Warrior Programs — Faithbased healing for veterans

https://www.mightyoaksprograms.org

Wounded Warrior Project — Resources, benefits, peer support

https://www.woundedwarriorproject.org

ResponderStrong — Mental wellness for first responders

https://responderstrong.org

Addiction & Recovery

Celebrate Recovery — Christcentered 12step program

https://www.celebraterecovery.com

SAMHSA Helpline — 24/7 mental health and addiction help

1800662HELP (4357)

https://www.samhsa.gov/findhelp/nationalhelpline

The Hope Line — Christian support for those in crisis

https://www.thehopeline.com

Suicide Prevention

National Suicide Prevention Lifeline — 988

https://988lifeline.org

Chapter 43

Matthew

Christ the King Who Heals the Broken and Fulfills the Promise

Chapter 43. Matthew

Christ the King Who Heals the Broken and Fulfills the Promise

"This is the genealogy of Jesus the Messiah the son of David, the son of Abraham…"

— Matthew 1:1

Matthew opens not with fireworks, but with a genealogy, a lineage of names, some revered and some scandalous. To us, it might look like a forgotten scroll of dusty history, but to the Jewish mind and to those of us seeking stability in a fractured world, it's a declaration: Jesus is the promised King, and His arrival changes everything.

The family tree in Matthew 1 isn't polished. It's filled with exiles, outcasts, and sinners, just like us. That's the point. Jesus stepped into the brokenness of a real family line, and He steps into ours too. His reign begins not with perfection, but with presence. For veterans, first responders, trauma survivors, and those in recovery, this is hope: you don't have to be perfect to belong to the King.

A King Who Brings Peace, Not Pressure

In a world that prizes power and control, Matthew's Gospel reveals a different kind of ruler. This King enters the scene humbly, born in a stable, raised among blue collar workers, baptized in a river filled with sinners, and when He begins to speak, He flips everything on its head:

"Blessed are the poor in spirit… those who mourn… the peacemakers."

— Matthew 5:3–9

These aren't the typical heroes. They're the crushed, the grieving, the worn out. In the Sermon on the Mount, Jesus lays out a kingdom built on mercy, not merit. For those of us who've spent years trying to prove ourselves whether in battlefields, boardrooms, or barstools, Jesus whispers something better: Come to Me, and I will give you rest.

"Come to Me, all who are weary and burdened… and you will find rest for your souls."

— Matthew 11:28–29

A Kingdom for the Traumatized and Tired

Matthew's Gospel doesn't ignore pain. It walks us through the raw moments of betrayal, fear, and death. But it also shows us Christ's steady rule in the midst of chaos.

If you've experienced trauma, flashbacks, addiction, panic attacks, or self-loathing, Matthew introduces a King who isn't scared of your wounds. He sees the veteran with night sweats, the addict fighting cravings, the survivor who can't stop shaking, and He says, You're mine, and I'm not letting go.

Jesus doesn't rule from a distance. He steps into the fire. He touches the leper. He eats with the outcasts. He calls fishermen, tax collectors, and zealots everyone from very kind of background, and turns their pain into purpose.

Fulfillment for the Forgotten

Over and over, Matthew writes: "This happened to fulfill what was spoken…" He draws a golden thread from Genesis to Jesus. For every broken promise you've experienced in this life government promises, military oaths, relational betrayals Christ fulfills what no one else could. "The virgin will conceive…" "Out of Egypt I called My Son…" "He will be called a Nazarene…" Christ is the proof that God never forgets His word. He doesn't abandon us in our silence, our addiction, or our depression. He shows up. Every time.

A King Crowned with Thorns

Matthew doesn't skip over the agony of the cross. He brings us face to face with betrayal, with abandonment, with the cruel mockery of soldiers and the scream of Jesus in the dark.

Why? Because he wants us to see that this King knows trauma. He carried the sin of the world. He sweat blood. He was deserted by His closest friends, and He didn't back down.

In Matthew 27, the King wears a crown of thorns, not to shame Him, but to declare forever: I understand pain. I understand injustice. I understand what it feels like to cry, "My God, why have You forsaken Me?"

Then, three days later, the tomb breaks open. Trauma doesn't win. Death doesn't win. Christ, the King, rises and invites us to rise with Him.

"All authority in heaven and on earth has been given to Me. Therefore go…" — Matthew 28:18

The King Who Restores

You may have served under commanders who failed you. You may have given years of your life to addiction, anger, fear, or codependency. You may have scars on your body or your soul, but there is a King who is not ashamed of you. A King who weeps with you, walks with you, and fights for your healing. A King who turns guilt into grace and pain into purpose. His name is Jesus, and His Kingdom is wide enough for you.

Lay down your weapons. Let go of the shame. Bow your heart, not in defeat, but in freedom. You don't have to earn peace. You just have to let the King give it to you.

The King Who Sends the Wounded

The Gospel of Matthew doesn't end with a retreat, it ends with a mission. After the cross, after the resurrection, after grief and disbelief, Jesus meets His disciples on a mountain and speaks these final words:

"All authority in heaven and on earth has been given to Me. Therefore go and make disciples…" — Matthew 28:18–19

Notice who He says this to: not perfect people, not unscarred saints, but to followers who had doubted, denied, and run. He speaks to the ones still recovering from their own failure and fear, and He sends them.

Why is this important? Because many veterans, trauma survivors, and people in recovery disqualify themselves from ever being used by God again. We think, "I've messed up too much. I'm too broken. I'm just trying to survive—how could I help someone else?" Jesus doesn't see you as disqualified, He sees you as commissioned.

Your scars don't sideline you; they become signs of hope to others still in the trenches. The King who wore thorns has chosen to send you, not in your strength, but in His.

So maybe your next step isn't to "do more," but to believe this: you're still part of the mission. You're still in the story, and the King still walks with you, even when your knees are shaky and your voice trembles.

He is with you always, in the fire, in the silence, and in the healing that follows.

Chapter 44

Mark

Christ the Servant King Who Gives His Life

Chapter 44. Mark

Christ the Servant King Who Gives His Life

"For even the Son of Man did not come to be served, but to serve, and to give His life as a ransom for many."

— Mark 10:45

Mark wastes no time.

Unlike Matthew or Luke, there's no Christmas story, no angels or genealogies, just action. "The beginning of the good news about Jesus the Messiah, the Son of God" (Mark 1:1), and we're off. In chapter one alone, Jesus is baptized, tested in the wilderness, announces the Kingdom, calls disciples, heals the sick, casts out demons, and rises before dawn to pray. It reads like a war journal, tight, tense, mission driven.

For those of us who have lived in combat zones or survived seasons of trauma, that urgency feels familiar. There's no time for long introductions when you're in survival mode. The world's on fire, and something has to be done now. That's how Mark writes, and that's how many of us have lived.

Woven through the pace is a powerful truth: Christ came not to be served, but to serve. He is the Servant King, not seated high above the mess but down in it with us healing, listening, lifting, feeding, forgiving, and, ultimately, laying down His life.

The Servant Who Moves Toward the Broken

Mark's Jesus is always on the move. He's not waiting for people to come to Him. He goes where the pain is.

He heals Peter's mother-in-law in a private home (Mark 1:29–31), touches a leper that everyone else avoided (Mark 1:40–45), forgives and raises a paralyzed man lowered through a roof (Mark 2:1–12), and speaks peace to a storm tossed sea (Mark 4:35–41). He meets people in graveyards, marketplaces, synagogues, and lonely places. And in every setting, the broken are seen.

As a veteran, I remember the feeling of being "unseen" when I got home. The world kept spinning, but I had changed. I carried wounds you couldn't see. My anger and anxiety boiled beneath the surface. I didn't know if God still wanted anything to do with me or if I even wanted anything to do with Him.

Then I encountered Jesus in the Gospel of Mark, and something clicked. He didn't avoid people like me, He walked right toward them. He wasn't afraid of trauma. He didn't shy away from demonic torment, mental anguish, bleeding women, or desperate fathers. He embraced them. Touched them. Listened to them. Healed them.

As Dallas Willard wrote, "The kingdom of God is not a matter of getting people into heaven, but of getting heaven into people." Mark's Jesus does just that, He brings heaven to us in the chaos of our now.

The Authority to Heal, The Compassion to Stay

One of the first things people noticed about Jesus was His authority. "He taught them as one who had authority, not as the teachers of the law" (Mark 1:22). What makes His authority beautiful is His compassion. Power without compassion is tyranny. Compassion without power is just sentiment. In Jesus, we see strength submitted to love.

When He looked out at the hungry crowds, Mark tells us, "He had compassion on them, because they were like sheep without a shepherd" (Mark 6:34). And then He fed them. He didn't just feel something, He did something.

For first responders and combat veterans, this tension between power and tenderness can feel real. We were trained to take action, to dominate chaos, but when the uniform comes off, it's often harder to connect with our own emotions. Some of us numbed our pain with alcohol or drugs. Some with rage. Others with silence. Here, in Jesus, is a man who is both lion and lamb. Fierce against evil, yet gentle with the broken, and through His Spirit, He teaches us how to do the same.

The Call to Sacrifice

Mark's Gospel is not for the fainthearted. Jesus doesn't sugarcoat the cost of following Him:

"Whoever wants to be My disciple must deny themselves and take up their cross and follow Me." — Mark 8:34

That sounds intense, and it is, but this isn't about suffering for the sake of suffering. It's about surrender. It's about giving up our lives of self-rule and stepping into something deeper, eternal. For those of us in recovery, from trauma, addiction, depression, or all three this hits home. Taking up your cross might mean surrendering control. Laying down the weapons of self-destruction. Breaking the cycle. Walking through grief without medicating. It might mean going back to the people you hurt and seeking reconciliation. That's not easy, but it's worth it. Jesus doesn't call us to anything He hasn't done Himself. He bore the cross first. He went to the darkest place first. So now, when we follow Him, we follow someone who has been there.

The Final Week: Our Suffering Servant

The final chapters of Mark zoom in on Jesus' last days. The betrayal, the trial, the mockery, and the cross. He is mocked as a fake king. Beaten. Stripped and nailed. He cries out, "My God, my God, why have You forsaken Me?" (Mark 15:34). If you've ever felt abandoned by God or felt the silence of heaven in the middle of your suffering, Jesus knows exactly what that feels like. The beauty is, He didn't stay in the silence. The tomb didn't win.

Mark ends his Gospel with an empty grave, and though the earliest manuscripts leave it open ended, maybe that's the point. The story isn't over. The resurrection isn't just Jesus' victory, it's the beginning of ours.

Veterans, Servants, and the Mission Forward

So where does that leave us, veterans, first responders, recovery warriors? Mark answers with Jesus' life: Serve. Love. Move toward the broken. Carry your cross. Live with purpose. Die to your pride. Rise in His peace.

We're not disqualified. Our past doesn't cancel our calling. Christ's sacrifice restores it. Our story doesn't end in addiction, in guilt, or in bitterness. It begins again at the empty tomb.

You are created to be both lion and lamb." In Jesus, we find the model and the power to become both. Not to fight meaningless wars, but to live with courageous compassion and eternal peace.

Chapter 45

Luke

Christ the Savior for All People

Chapter 45. Luke

Christ the Savior for All People

"For the Son of Man came to seek and to save the lost."

— Luke 19:10

Luke's Gospel is the longest of the four, and in many ways, the most detailed and intimate. As a physician, Luke understood pain, the human condition, and the power of healing. As a companion of Paul, he carried the heart of a missionary, an urgency that every soul matters. That's what we feel in every chapter, Jesus didn't come just for the religious or the righteous, He came for the wounded, the weary, and the wanderer.

Veterans. First responders. Those in addiction recovery. Those who have walked through trauma and felt forgotten by the world. Luke is your Gospel.

This isn't just a story about God, it's the story of a God who stepped into the dirt, dust, and despair of humanity to pull us into redemption. A Christ who came not for the clean and polished, but for the broken and burdened, and if that's you you're in the right place.

The Humanity of Christ

Luke is the only one who gives us the shepherds in the field. While Matthew gives us wise men and kings, Luke gives us the poor and unknown. Why? Because Jesus didn't just come to save humanity, He came to be human. To live it. Feel it. Bleed in it.

He came for: the poor (Luke 6:20–21), the prisoner (Luke 4:18), the overlooked woman (Luke 8:43–48), the abused and abandoned child (Luke 18:16), the shamed addict, the disgraced soldier, the exiled soul

When I first started seeking Christ, I didn't feel worthy to be found. I was still knee deep in addiction. Still fighting demons from combat. Still hurting people who loved me, but Luke's Jesus didn't flinch. He met me in my mess and didn't walk away.

J.I. Packer once wrote: "What matters supremely… is not the fact that I know God, but the larger fact which underlies it, that He knows me." Luke teaches us that God doesn't just see us from a distance, He knows our name, our wounds, and our worth.

Christ the Friend of Sinners

One of Luke's deepest themes is how Jesus responds to sinners and it's not with condemnation. It's with compassion. He called Levi, the tax collector, right from the job site (Luke 5:27–32). He stopped for Zacchaeus, the most hated man in town (Luke 19:1–10). He told stories of lost sheep, lost coins, and lost sons—and every story ended in celebration, not rejection (Luke 15). He was anointed by a woman with a shameful past who didn't belong at the table, but He made her the center of the moment (Luke 7:36–50).

For any of us carrying guilt over what we've done, what we've seen, or what we've survived, Luke offers hope. Not a cheap kind, but the kind that looks you in the eye and says, "You still belong. You're not too far gone. You are not disqualified."

For those in recovery, whether it's from trauma, addiction, or years of silence Luke reminds us: Christ does not recoil from your story. He redeems it.

Christ the Healer

It's no surprise that the physician writes most about healing.

He heals ten lepers but only one returns to say thank you (Luke 17:11–19). He restores the eyes of a blind man (Luke 18:35–43). He gives legs to the paralyzed (Luke 5:18–26). He casts out demons and restores sanity to the tormented (Luke 8:26–39). We often carry wounds no one can see. The kind you don't show on VA intake forms. Luke's Jesus sees them. He doesn't turn away from our pain, He walks right into it.

Pete Walker, in Complex PTS, writes: "The child does not die, but is banished into the body, where it waits for safe enough conditions to reemerge and be cared for." Veterans and trauma survivors know this reality. Our inner pain doesn't vanish. It hides until the Healer calls it forth. Jesus is the One who provides those safe conditions. In Him, we're safe enough to feel again, grieve again, and begin to heal.

The Cross: Mercy in the Middle of Suffering

No one records Jesus' final moments quite like Luke.

He prays for His executioners: "Father, forgive them." (Luke 23:34), He reassures the thief dying next to Him: "Today you will be with Me in paradise." (Luke 23:43), He gives up His spirit in peace, not rage: "Into Your hands I commit My spirit." (Luke 23:46), even on the cross, Jesus is still saving. Still healing. Still loving.

For those of us who have faced the battlefield physical or spiritual this matters. We've seen enough brutality. We've participated in it. We've hurt and been hurt, but Luke's Christ is grace on a cross. The message is clear: There is no wound too deep for the love of God.

The Resurrected Christ: Peace That Stands in Fear

When Jesus rises, His first words are: "Peace be with you." (Luke 24:36)

He shows them His scars. Eats with them. Walks beside them. Opens the Scriptures. In other words He doesn't just appear; He stays.

For combat veterans and trauma survivors, this peace isn't sentimental. It's survival. We know what it means to long for peace in our minds, peace in our memories, peace in our sleep. Jesus offers a peace that doesn't vanish when the anxiety hits. His peace is presence, and His presence never leaves.

John Eldredge writes: "The peace of God is not the absence of trouble but the presence of Jesus in the trouble." That's Luke's final word: Peace is not found in perfect circumstances. Peace is found in the Person of Christ.

Chapter 46

John

Christ the Word Made Flesh

Chapter 46. John

Christ the Word Made Flesh

"In the beginning was the Word, and the Word was with God, and the Word was God."

— John 1:1 (NIV)

If the other Gospels show us Jesus in history, John shows us Jesus in eternity. From the very first verse, we're pulled into a reality deeper than time: the eternal Word, the Son of God, who stepped into flesh and walked among us. This is not just theology, it's hope wrapped in skin.

John doesn't open with a manger or a genealogy. He opens with glory. "The Word became flesh and made his dwelling among us..." (John 1:14). Not just visiting. Dwelling. Camping out in our chaos.

Christ in the Dirt

Jesus doesn't shout from a distance, He gets close. He walks roads with blisters. He cries at tombs. He touches the outcasts. He bleeds. He sweats. He dies, and when He rises, He still carries the scars. For the battle worn and broken, this changes everything. The One who made galaxies also made time to restore the soul of a wounded man by a pool (John 5), a woman at a well (John 4), and a doubter named Thomas (John 20).

"I Am": The God Who Meets You

John's Gospel is built around seven "I Am" statements, echoes of God's name revealed in Exodus 3:14. Each one is a lifeline for the hurting.

> "I am the Bread of Life" – He feeds your soul in famine.
>
> "I am the Light of the World" – He shines into the darkest corners of addiction and depression.
>
> "I am the Gate" – He guards your soul from the enemy's lies.
>
> "I am the Good Shepherd" – He carries the wounded and seeks the lost.
>
> "I am the Resurrection and the Life" – He defeats death and despair.
>
> "I am the Way, the Truth, and the Life" – He's not a map—He's the journey.
>
> "I am the True Vine" – He gives life to those who've been cut off and dried out.

Each "I Am" is a whisper to your weary heart: "I'm here. I'm enough. I'm not leaving."

Conversations That Heal

John doesn't just tell stories he invites us into holy conversations:

> With Nicodemus in the dark: "You must be born again." (John 3)
>
> With the woman at the well: "I know your past… and I still offer you living water." (John 4)
>
> With the man born blind: "This happened so that the works of God might be displayed in him." (John 9)

> With Martha, grieving her brother: "I am the resurrection and the life." (John 11)

Jesus doesn't run from pain, He walks right into it. He doesn't avoid hard questions, He answers with grace and fire. He doesn't flinch at your sin, He offers water, healing, and Himself.

The Scars That Speak

When He rises, Jesus doesn't hide His scars. He shows them to Thomas. He cooks breakfast for Peter. He calls Mary by name through her tears. In John's Gospel, the resurrected Christ is relentlessly personal, and He still is. You are not who others say you are. You are not who your trauma says you are. You are who God says you are. In John, God says: You are loved. You are chosen. You are His.

Veterans and John's Gospel

John's Gospel doesn't ignore trauma, it brings it into the light. Jesus knows what it means to weep, to bleed, to carry pain, and to love through suffering. Your scars don't disqualify you. His scars qualify you. Your mission didn't end with your service. You're still called. Your identity isn't lost it's hidden in Christ. When Jesus breathes on His disciples and says, "Peace be with you… As the Father has sent Me, so I am sending you" (John 20:21), He includes you.

Chapter 47

Acts

Christ the Risen Lord and Spirit Filled Commander

Chapter 47. Acts

Christ the Risen Lord and Spirit Filled Commander

"But you will receive power when the Holy Spirit comes on you; and you will be my witnesses in Jerusalem, and in all Judea and Samaria, and to the ends of the earth." — Acts 1:8

If the Gospels are the boots on the ground life of Christ, Acts is the deployment of His mission. The war for souls didn't end at the empty tomb. It intensified. Acts is not the end of the story, it's the beginning of a movement that's still marching today.

Christ may have ascended, but His presence never left. He gave His Spirit, and with it, a command: Go. For those of us who served, veterans, first responders, and anyone who's endured trauma, Acts reads like a spiritual after action report. It's filled with sacrifice, calling, courage, community, persecution, and purpose. This isn't a quiet history lesson. It's a call to arms.

Christ, Our Ascended Commander

Acts opens with Jesus ascending, not abandoning. His exit wasn't retreat. It was enthronement. He didn't leave us leaderless, He left us armed with the Spirit. That matters to someone like me. I spent years under broken authority and then more years bristling under some good and some not so good authority in the Army. I've had leaders I trusted with my life and others I wouldn't follow across the street.

But Jesus? I trust Him. He doesn't send us into a fight He hasn't already won. He doesn't shout orders from a distance. He equips us with Himself through the Holy Spirit.

Pentecost: Heaven's Fire and Wind

In Acts 2, the Holy Spirit shows up, not as a whisper, but a roar. Tongues of fire. Rushing wind. A roomful of broken men transformed into bold messengers.

I remember when we got our first orders before Afghanistan. There was this adrenaline rush the weight of the gear, the roar of the engines, the look in each soldier's eyes. We knew it was real.

Pentecost was heaven's deployment moment. The church was no longer a hiding place it became an outpost. The mission had begun. and that mission still stands. If you've ever wondered what your life is for, if you've lost your purpose after taking off the uniform or coming out of the storm, look to Acts. You've already been given your orders: Go. Testify. Make disciples. Love boldly.

Brotherhood and Belonging

Acts 2:42–47 describes a group of believers doing life together breaking bread, praying, praising, sharing all things in common. This wasn't kumbaya Christianity. It was forged in the fire fellowship. That's what I miss about the military most. Brotherhood. Camaraderie. People who had your six. You didn't have to be perfect, you just had to show up, lock in, and keep each other alive.

The early church had that same raw unity. They weren't just church members they were a spiritual combat unit. One mission. One Spirit. One Lord.

For those of us in recovery, this also hits deep. Whether you've come out of addiction, trauma, or shame, there's power in locking arms with others who know the battlefield. In Acts, no one heals alone. Neither should we.

Boldness in the Face of Fire

Acts isn't all miracles and mountaintops. It's prison cells. Martyrdom. Beatings. Betrayals. It's spiritual war, and yet, again and again, the Spirit gives boldness.

Acts 4:13 says, "When they saw the courage of Peter and John... they were astonished and took note that these men had been with Jesus."

That verse sticks with me. I've seen combat courage. But spiritual courage that takes a different kind of guts. Courage to forgive. To stand for truth. To say "no" when the old life calls your name. To walk into a church after years of running. You don't need to be fearless. You need to be Spirit filled. That's what gives us boldness, not bravado. Not perfection. Just presence.

Stephen: The First to Fall

Stephen, full of grace and power, preaches the truth and pays with his life. As the stones fly, he sees Christ standing not sitting at the right hand of God. Everywhere else in Scripture, Jesus is seated in glory, but when Stephen falls, the Savior stands. That image moves me. I think of my brothers who never made it home. The ones whose names are written in my heart. The ones who died by enemy hands, by overdose, by suicide, or by silence. Acts tells me: Christ stood for them too. He stands for the hurting. He doesn't look away. He stands with us and for us.

From Terrorist to Testifier

Saul the enemy of the church is confronted by Christ Himself and transformed into Paul, the apostle to the nations. This is my favorite story in Acts, maybe in the whole Bible. Because I was Saul too. Running from God. Hurting others. Hardhearted and high. I was a wreck wrapped in skin but Jesus didn't write me off. He met me in my mess. No one is too far gone. No addiction too deep. No trauma too twisted. The Damascus Road is still open.

Paul's Journey: Perseverance With a Purpose

Paul's journeys take him through storms, stonings, shipwrecks, snakebites, and imprisonment but the Gospel moves. It never stops. Acts 28 doesn't end with a period it ends mid thought, because the mission isn't over. We are Acts 29. The Spirit still moves. The mission still matters.

Like Paul, we may walk with a limp. We may bear scars, but we walk forward. We speak truth. We love well. We carry the name of Christ to the broken and the bound.

Christ in Acts for Warriors and Wounded

Christ the Commander He gives orders with love and authority.

Christ the Presence — His Spirit lives in us.

Christ the Redeemer — He rewrites our stories.

Christ the Strength — He gives courage to endure.

Christ the Mission — He sends us with purpose and power.

No matter what battlefield you've walked through combat, addiction, grief, or mental illness Acts reminds you: You still have a mission.

Chapter 48

Romans

Christ, Our Righteousness and Peace with God

Chapter 48. Romans

Christ, Our Righteousness and Peace with God

"Therefore, since we have been justified through faith, we have peace with God through our Lord Jesus Christ."

— Romans 5:1

Romans is the battlefield map of the Gospel. It is Paul's strategic letter to both Jews and Gentiles, laying out the mission plan of salvation, the terrain of sin, the stronghold of grace, and the rallying cry of righteousness in Christ.

Where the Gospels introduce us to Jesus and Acts sends us on mission, Romans explains why we need the Gospel in the first place, and how it changes everything. It doesn't sugarcoat the human condition. It exposes the raw, wounded truth of who we are without Christ, but it also raises the banner of hope: righteousness is not achieved; it is received.

For the veteran who has been through fire, for the addict, the angry, and the ashamed Romans declares a ceasefire between you and heaven. Not because you surrendered first, but because Christ did.

The Problem and the Rescue Mission

Romans begins with hard truth: "All have sinned and fall short of the glory of God." — Romans 3:23

There's no "us" and "them" here. No moral high ground. Paul says we've all failed inspection. Whether you've worn a uniform or a jumpsuit, whether your sins were on display or buried deep every one of us is guilty. Paul doesn't leave us there. In the very next breath, he gives the remedy:

"...and all are justified freely by His grace through the redemption that came by Christ Jesus." — Romans 3:24

That's courtroom language, but it's also combat language. In military terms, we were cut off behind enemy lines, surrounded by sin, low on hope, unable to save ourselves. Then Christ breached the line. He took our place. He declared us justified, not because we earned it, but because He paid it.

This isn't just about forgiveness, it's about freedom. The righteousness of God is not a concept; it's a covering. A new identity. A new uniform, stained not with our sin but with His blood.

Peace with God: The End of Hostilities

"Since we have been justified through faith, we have peace with God..." — Romans 5:1

For many veterans, the fight doesn't end when the deployment does. The war follows you home in memories, in relationships, in sleepless nights. Romans speaks directly to that kind of soul deep unrest. Peace with God means more than just "not fighting." It means no more condemnation, no more wondering if you've done enough, and no more trying to prove yourself to heaven. Peace with God means the Judge has declared you innocent, the Commander has called you righteous, and the Father has called you home.

You don't need to keep punishing yourself. The sentence has been served. The war is over. The banner over your life now reads "grace."

Grace Greater Than Guilt

Romans 6 reminds us that grace is not permission to sin it's power to overcome it. "We were therefore buried with him through baptism into death... so that... we too may live a new life." — Romans 6:4 The old life? Dead. The old identity? Buried. The addictions, the anger, the shame, the violence, the failures all nailed to the cross with Christ. When He rose, so did we. Resurrection power doesn't just change our future, it transforms our present.

Romans 7 is raw. Paul confesses the inner war we all know too well: "I do not do the good I want to do, but the evil I do not want to do—this I keep on doing." — Romans 7:19. Romans 8 is the answer to that agony: "There is now no condemnation for those who are in Christ Jesus." — Romans 8:1. Read that again: no condemnation. Not reduced charges. Not a lighter sentence. None. Zip. Zero. This is the oxygen of the Gospel. This is the fresh air your soul has been gasping for. Not because you cleaned up, but because Christ covered you.

Nothing Can Separate Us

Paul crescendos in Romans 8 with what may be the most powerful promise in Scripture:

"I am convinced that neither death nor life, neither angels nor demons... nor anything else in all creation, will be able to separate us from the love of God..." — Romans 8:38–39

Not your past. Not your divorce. Not the bottle.

Not the war. Not the trauma. Not even your unbelief.

God's love doesn't let go. You can rage, run, and wrestle, but you cannot out run His mercy.

Identity, Not Performance

Romans chapters 9–11 deal with God's sovereign plan, and chapters 12–16 get intensely practical: love sincerely, serve faithfully, bless enemies, overcome evil with good.

This is where righteousness becomes real life. You're not just saved from something, you're saved for something. Romans 12:1 calls us to present our lives as living sacrifices not as payment, but as praise. You don't have to fake strength anymore. You don't have to wear the mask. In Christ, your true identity is secure, and your mission continues.

For the Veteran

Romans is for:

The warrior with invisible wounds, the addict who swore they'd never relapse, the father who walked out and can't forgive himself, the woman who thinks she's too far gone, the man who served with honor but feels hollow inside, Romans says: You are justified. You are at peace. You are deeply loved. You are not condemned. You are more than a conqueror in Christ.

Chapter 49

1st Corinthians

Christ, the Wisdom and Power of God

Chapter 49. 1st Corinthians

Christ the Wisdom and Power of God

"For the message of the cross is foolishness to those who are perishing, but to us who are being saved it is the power of God."

— 1 Corinthians 1:18

A Church in Conflict, a Cross That Unites

Corinth was a spiritual war zone. The church there was fractured by divisions, pride, sexual sin, lawsuits, and confusion about spiritual gifts. It was a mess and Paul doesn't sugarcoat that. Instead of condemning them or walking away, he brings them back to the center:

"For I resolved to know nothing while I was with you except Jesus Christ and Him crucified." — 1 Corinthians 2:2

In a world obsessed with status, knowledge, and strength, Paul pointed them to the cross. Not a symbol of human wisdom, but of divine power. Not a trophy of achievement, but a battlefield where victory was won by surrender.

Veterans understand division. We've seen brotherhood turn to bitterness. Sometimes the hardest battles are not fought overseas but inside our homes, churches, and minds. Paul reminds us: only the cross can unify what the flesh divides.

The Veteran's Struggle and the Power of the Cross

We often try to fix our pain with the tools we know grit, control, withdrawal, and distraction, but you can't outmuscle a spiritual wound. You can't medicate a soul with strategy, and you can't heal trauma by hiding it. The message of the cross is that healing starts with surrender. Jesus took our chaos and absorbed it in His body. He met our violence with forgiveness, our guilt with grace. The battlefield taught us to survive. The cross teaches us how to live.

The Body of Christ: Many Parts, One Mission

Paul uses a powerful metaphor to describe the Church:

"Now you are the body of Christ, and each one of you is a part of it." — 1 Corinthians 12:27

Veterans understand this intuitively. Every role matters. Every soldier, sailor, airman, or marine has a purpose. The unit doesn't function if one piece fails. The same is true in the Church. You may feel disqualified because of trauma, addiction, failure, or shame, but Paul says you still belong. Your scars don't exclude you they're part of your testimony. You may not serve in the same way you once did, but you are still vital to the mission. There are no unnecessary Christians.

Love: The Greatest Weapon

1 Corinthians 13 is more than a wedding reading, it's a call to arms for spiritual warfare:

"And now these three remain: faith, hope and love. But the greatest of these is love." — 1 Corinthians 13:13

Love isn't weakness. It's not passive. It's the strongest weapon we have. Love endures. Love bears all things. Love outlasts trauma and outshines bitterness. On the battlefield, we're taught to harden ourselves. In Christ, we are called to soften not in passivity, but in redemptive power. We love because He first loved us. We forgive because He forgave us. We serve because He served us unto death.

The Resurrection: Death Has Lost Its Sting

Paul anchors the entire letter in resurrection hope:

"Death has been swallowed up in victory. Where, O death, is your victory? Where, O death, is your sting?" — 1 Corinthians 15:54–55

Veterans often carry the weight of death, friends lost, near death experiences, and survivor's guilt, but Paul's declaration is loud and clear: death does not win. Christ has conquered the grave and because He lives, we will too. This isn't fantasy. It's future fact, and it changes how we live today, with courage, purpose, and peace.

For the Veteran

1 Corinthians reminds us: You don't have to have it all together. Corinth didn't. Christ still showed up. Your wounds don't make you unusable. You're part of the body, by grace. Love is your mission. Not performance, not pride. Love. The cross is your anchor. Not your record. Not your strength. Christ crucified. The resurrection is your hope. Not just for heaven, but for today's battles. You are not too broken. You are not forgotten. You are not alone.

Chapter 50

2nd Corinthians

Christ, Our Comfort in Affliction

Chapter 50. 2nd Corinthians

Christ, our Comfort in Affliction

"Praise be to the God and Father of our Lord Jesus Christ, the Father of compassion and the God of all comfort, who comforts us in all our troubles, so that we can comfort those in any trouble with the comfort we ourselves receive from God." — 2 Corinthians 1:3–4

A Letter from the Frontlines of Suffering

This isn't Paul the polished theologian. This is Paul, the wounded warrior of the Gospel, writing from a place of raw humanity. He doesn't hide the weight of his trials. He talks about despair that felt like death, burdens too heavy to carry, and weakness that left him fully dependent on Christ.

"We were under great pressure, far beyond our ability to endure, so that we despaired of life itself." — 2 Corinthians 1:8

If you've ever felt like you were at your end emotionally, mentally, or physically this letter is for you. It's not written from a pulpit; it's written from the trenches, and in those trenches, Paul points to the One who never left his side: Christ, the God of all comfort.

Strength Revealed in Weakness

Paul pleads with God to remove a "thorn in the flesh," some torment that humbled him, but God's answer is staggering:

"My grace is sufficient for you, for my power is made perfect in weakness." — 2 Corinthians 12:9

The world says hide your weakness. God says, That's where My glory shines. Veterans and first responders are taught to appear strong, composed, in control. But what if God's power isn't found in pretending? What if it's found in trusting right in the middle of the collapse? Your lowest moment might be the very place where Christ shows up the loudest.

The Ministry of Reconciliation

One of the clearest Gospel truths in Scripture lives in this letter:

"God was reconciling the world to Himself in Christ, not counting people's sins against them… And He has committed to us the message of reconciliation." — 2 Corinthians 5:19

The war is over. God made peace with us through the cross, and now we carry that peace into every relationship whether it's a strained marriage, a lost friendship, or an inner war between shame and grace. You don't have to be perfect to be a peacemaker. You just have to be forgiven.

The Glory That Outweighs the Pain

Paul doesn't downplay suffering. He elevates the glory that follows it:

"Therefore, we do not lose heart. Though outwardly we are wasting away, inwardly we are being renewed day by day… For our light and momentary troubles are achieving for us an eternal glory that far outweighs them all."

— 2 Corinthians 4:16–17

That's not denial, it's defiance. Defying the lie that this life is all there is. Defying the darkness that whispers, "It will never get better." Glory is coming. Peace is coming. Full healing is coming, and until then, Christ walks with us through every fire.

For the Wounded Warrior

2 Corinthians reminds us: You are not disqualified by your weakness. You are qualified by His grace. You don't walk alone. The Comforter goes with you and so do those He's redeemed. Your scars have a purpose. They're not shameful they're sacred. The Gospel isn't just about forgiveness. It's about reconciliation, restoration, and renewal. Eternal glory awaits. Your suffering is not the end.

Chapter 51

Galatians

Christ, Our Freedom and Identity

Chapter 51. Galatians

Christ Our Freedom and Identity

"It is for freedom that Christ has set us free. Stand firm, then, and do not let yourselves be burdened again by a yoke of slavery."

— Galatians 5:1

Breaking the Chains

Galatians is Paul's bold and urgent call to reclaim freedom not political freedom, but spiritual freedom. Paul is furious that the Galatian believers are turning back to religious performance, legalism, and fear. He doesn't sugarcoat anything. They had been set free by Christ, but now they were being seduced into earning God's love all over again.

For anyone who's been trapped by addiction, haunted by the past, or bound by shame, Galatians is a rescue flare in the dark. It declares what many of us didn't believe for far too long: you don't have to work for what God gives freely.

"You foolish Galatians! Who has bewitched you?" — Galatians 3:1

Sometimes we need truth to hit like a slap in the face. Because many of us veterans, first responders, and trauma survivors walk around in silent prisons. The prison of not feeling good enough. The prison of trying to make up for the damage we've done. The prison of performance based worth. Paul says: Christ didn't die to make you a better slave. He died to set you free.

Adopted. Redeemed. Known.

"But when the set time had fully come, God sent his Son… to redeem those under the law, that we might receive adoption to sonship." — Galatians 4:4–5

The word "adoption" is loaded with meaning. In the ancient world, adoption gave someone full legal rights as a family member. No second tier status. No reminders of your past.

"So you are no longer a slave, but God's child; and since you are his child, God has made you also an heir." — Galatians 4:7

That's the message so many of us long to hear: You're not a failure. You're not a mistake. You're not forgotten. You're family. I spent years trying to define myself by war, by wounds, by addiction, by loss. In Christ, I found a new name. A new identity. Not as a victim, not as a screwup, but as a redeemed son of God.

Freedom Is Not an Excuse

Paul is crystal clear: freedom in Christ is not a license to sin it's the power to finally live.

"You, my brothers and sisters, were called to be free, but do not use your freedom to indulge the flesh… walk by the Spirit." — Galatians 5:13,16

Freedom doesn't mean lawlessness. It means transformation.

When Christ sets us free, He also sends His Spirit to form His character in us not through religious rule following, but through inner renewal. That's what Paul describes as the fruit of the Spirit:

"Love, joy, peace, patience, kindness, goodness, faithfulness, gentleness, and self-control." — Galatians 5:22–23

These aren't things we fake they're things He grows in us.

For Those Who Keep Falling

"If someone is caught in a sin, you who live by the Spirit should restore that person gently… Carry each other's burdens, and in this way you will fulfill the law of Christ."

— Galatians 6:1–2

This is for the one who relapsed last night. For the one who lashed out again. For the one who still has nightmares and doesn't want to talk about it. Paul doesn't say "condemn them." He says "restore them." Gently. With compassion. Because all of us have needed that grace.

If you're walking through recovery, or trying to hold your family together, or feel like your faith is barely hanging on remember this: grace doesn't let go. Even when you fall, Christ doesn't walk away.

Chapter 52

Ephesians

Christ, Our Peace and Purpose

Chapter 52. Ephesians

Christ, Our Peace and Purpose

"But now in Christ Jesus you who once were far away have been brought near by the blood of Christ. For He Himself is our peace." — Ephesians 2:13–14a

A Blueprint for the Redeemed Life

Ephesians reads like a battle plan, not for war, but for restoration. It's less about fighting enemies and more about discovering who we really are in Christ. Paul writes with the clarity of a general giving orders and the compassion of a pastor offering hope. He reminds us that before anything else, we are chosen, adopted, redeemed, and sealed by the Holy Spirit. That identity isn't earned by combat medals, church attendance, or clean records it's secured by grace.

"For it is by grace you have been saved, through faith, and this is not from yourselves, it is the gift of God." — Ephesians 2:8

For the veteran or first responder who has been defined by titles, trauma, or regrets, this is a radical declaration: Your identity is not what you've done. It's who Christ says you are.

Peace Through Reconciliation

Paul's letter is saturated with reconciliation. Not just vertical between us and God, but horizontal between people who once stood worlds apart.

"He has destroyed the barrier, the dividing wall of hostility…" — Ephesians 2:14b

That line hits different for anyone who's felt far off. Whether because of sin, addiction, shame, or simply the invisible wall between civilians and combat veterans Christ tears it down. For those who've lived with survivor's guilt or isolation, this truth matters: you're no longer a stranger, you're family. The war between you and God is over. Peace has a name. His name is Jesus.

From Death to Life

Paul paints a stark contrast in chapter 2:

"As for you, you were dead in your transgressions and sins... but because of His great love for us, God, who is rich in mercy, made us alive with Christ..." — Ephesians 2:1,4–5

We weren't limping. We weren't wounded. We were dead until Christ revived us. That's the power of grace. It's not a crutch for the weak; it's resurrection for the dead. For the alcoholic who swore they'd never drink again but did. For the combat vet who doesn't even recognize the person in the mirror. For the child of trauma still haunted by screams in the dark. Christ makes you alive again.

"Even when I was too broken to reach for God, He reached for me."

Unity in the Mission

Ephesians 4 calls the Church to unity—not uniformity, but Spirit led teamwork:

"Make every effort to keep the unity of the Spirit through the bond of peace." — Ephesians 4:3

Veterans know the power of a unified team. One weak link in a fire team puts everyone at risk, but one strong bond holds the whole thing together.

The same is true in the Church. You're not called to go it alone. Isolation might feel safer, but freedom is found in fellowship. Whether in a recovery group, a men's ministry, or a kitchen table Bible study healing happens in community.

Walking Worthy in Christ

"Live a life worthy of the calling you have received."

— Ephesians 4:1

This isn't about proving your worth it's about living from it. We don't earn God's love, but we respond to it. Paul urges believers to walk in humility, gentleness, patience, and love. These aren't soft words they're weapons of grace in a world of hostility. For warriors coming home, these virtues aren't weakness they're the signs of Spirit forged strength.

Chapter 53 The Armor of God

Christ, Our Defender and Divine Equipment

Chapter 53. The Armor of God

Christ Our Defender and Devine Equipment

"Put on the full armor of God, so that you can take your stand against the devil's schemes." — Ephesians 6:11

Boots on the Ground

Every soldier remembers their first time lacing up combat boots fresh, tight, unfamiliar. There's a moment of anticipation and fear before stepping into a new kind of battle. In training, we were taught to never go out without being fully equipped: helmet, vest, weapon, ammo, radio, water, tourniquet. We were told it could be the difference between life and death, but the most dangerous battles aren't always in foreign countries. Some of the fiercest wars happen on home soil in marriages, in the mind, in the silence of sleepless nights. Trauma, addiction, depression, and shame, they don't care about borders or flags, and we've all known warriors who fought bravely overseas only to come home and lose the battle within.

That's why Paul's message to the Ephesians hits home. He speaks like a veteran commander to a weary battalion, giving them more than a pep talk. He gives them gear, spiritual armor forged not in steel, but in the very person of Christ.

The War Within and Without

"For our struggle is not against flesh and blood, but against... the spiritual forces of evil in the heavenly realms." — Ephesians 6:12

This isn't a metaphor for personal discipline. It's a declaration of war. There's a real enemy. Not just trauma, not just addiction, not just bad habits but a spiritual enemy who targets the wounded, the isolated, the battle worn. Veterans and first responders understand this. You don't need to see a uniform to know someone's in combat. You see it in their eyes, in their posture, in their silence, and Paul's warning is clear: You need armor. Not just once, not just on Sundays. Daily. Deliberately. Devotionally. This armor isn't something you earn. It's not self-improvement. It's spiritual equipment, gifted by grace, forged in the victory of Christ, and fitted for you in the heat of your own story.

1. The Belt of Truth

"Stand firm then, with the belt of truth buckled around your waist..."— Ephesians 6:14a

The belt wasn't for fashion. It held the rest of the armor together. Without it, everything fell apart. Truth is what keeps your soul intact when lies try to unravel it. The enemy whispers: You'll never change. You're too dirty. You've gone too far. But Jesus says: You are mine. You are forgiven. You are new.

Veterans often carry lies deep in the gut. Lies from war. Lies from broken families. Lies from relapse. The belt of truth reminds us who we are not because we earned it, but because Christ declared it.

2. The Breastplate of Righteousness

"...with the breastplate of righteousness in place..."

— Ephesians 6:14b

The breastplate guards the heart. For a soldier, that means life or death. For the believer, it means freedom from guilt and shame. But this isn't our righteousness. It's Christ's. The same Jesus who died on the cross for our sin now stands as our covering. When the enemy accuses you, you point to your chest and say, not guilty. Not today. Christ is my defense.

3. Feet Fitted with the Readiness of the Gospel of Peace

"...and with your feet fitted with the readiness that comes from the gospel of peace."

— Ephesians 6:15

Boots matter. In combat, one blister can take you out, and in spiritual warfare, how you walk matters too. Paul says your footing is the Gospel of Peace. That sounds strange in a war zone, peace? He is not talking about passive peace. It's peace won at a cost. Peace that walks into chaos and plants the flag of the Kingdom.

For those of us in recovery or returning from trauma, this means we don't just survive, we advance. We carry peace into prisons, rehabs, homes, and hurting hearts. Your scars become your credentials. Your limp becomes your mission.

"God does not call the qualified; He qualifies the called."

4. The Shield of Faith

"Take up the shield of faith, with which you can extinguish all the flaming arrows of the evil one."

— Ephesians 6:16

Faith isn't just believing when things are good. Faith is holding the line when the sky is falling. Paul describes flaming arrows not little doubts, but explosive lies that aim to burn everything you love.

The shield of faith isn't thin optimism, it's anchored trust in a God who doesn't change when your circumstances do. Some days, your faith will feel weak. That's okay. A shaking hand can still lift a shield. Faith isn't about strength it's about direction. Point it at Christ, and stand your ground.

5. The Helmet of Salvation

"Take the helmet of salvation…"

— Ephesians 6:17a

The helmet protects your head, your mind, your identity, and your thoughts. Some of us have taken spiritual shrapnel to the brain. Intrusive thoughts. Relapse regret. Flashbacks. Suicidal ideation and lies that play like a broken record. The helmet of salvation reminds you that your mind is not the enemy's playground. You belong to Christ. You are saved. You are secure. When the mental fog rolls in, strap on this truth: He who began a good work in you will carry it on to completion. (Philippians 1:6)

6. The Sword of the Spirit

"…and the sword of the Spirit, which is the word of God."

— Ephesians 6:17b

Everything else is defensive. This is your weapon, and it's not carnal, it's the Word of God.

Jesus used Scripture to defeat temptation and so should we. When shame comes, we quote Romans 8:1: "There is now no condemnation…" When fear comes, we quote Isaiah 41:10: "Do not fear, for I am with you…" When addiction claws at the door, we quote 1 Corinthians 10:13.

This sword is not just for pastors or theologians. It's for every believer in battle. Study it. Memorize it. Speak it out loud like a battle cry.

Praying Always: The Invisible Advantage

"And pray in the Spirit on all occasions with all kinds of prayers and requests."

— Ephesians 6:18

Paul doesn't end the armor list with a period, but with a call to pray. Because even with armor, a soldier needs command with communication. Prayer is our radio. Our lifeline and our call for backup.

In this war, we don't fight alone. We are part of a unit, the Body of Christ. We cover each other. We call out targets. We drag the wounded and we speak life over dry bones.

Veterans, Warriors, and the Armor of God

If you've worn body armor before, you know it doesn't make you bulletproof, but it does give you a fighting chance. The same is true with God's armor. Life will still hit hard. Trauma will still try to sneak in. Addiction may crouch at the door, but you are not defenseless. You are equipped by the Commander of Heaven's Armies. Christ, Himself, is your armor.

"The Lord is a warrior; the Lord is His name." — Exodus 15:3

Chapter 54

Philippians

Christ Our Joy and Contentment

Chapter 54. Philippians

Christ, Our Joy and Contentment

"Rejoice in the Lord always. I will say it again: Rejoice!"

— Philippians 4:4

Joy in the Middle of the War

Paul wrote Philippians from prison, not a place you'd expect to find joy. Yet the letter radiates with it. Over and over, Paul speaks of rejoicing, peace, and contentment, not because life is easy, but because Christ is near.

For many of us veterans, first responders, those recovering from addiction, or carrying trauma, joy can feel like a distant memory. We've seen too much, lost too much, and fought too long. Yet Paul's words remind us: joy isn't rooted in comfort, it's rooted in Christ.

"The joy of the Lord is not the absence of pain; it's the presence of a Savior who never leaves your side."

The Fellowship of the Broken

"In all my prayers for all of you, I always pray with joy because of your partnership in the gospel…"

— Philippians 1:4–5

Paul's deep affection for the Philippians was forged through suffering. They had supported him, suffered with him, and stood beside him. That's the kind of fellowship those of us with military or recovery backgrounds recognize a brotherhood in the trenches. You don't bond through comfort, you bond through combat. The church in Philippi was that kind of church. They were "partners in grace," not just hearers of the Word but warriors beside Paul, and Paul knew: suffering doesn't weaken your faith, it welds it to others.

Peace That Guards the Mind

"Do not be anxious about anything, but in every situation… present your requests to God. And the peace of God… will guard your hearts and your minds in Christ Jesus."

— Philippians 4:6–7

The Greek word for "guard" here is a military term. Paul is saying that God's peace sets up camp around your heart like sentries guarding a fortress. He doesn't promise to remove the battle, but He promises to defend you in it. This is combat level peace, peace that holds even when memories flood in or triggers fire unexpectedly. It's not passive; it's active. It fights for you, and it's stronger than any darkness.

Christ Is Enough

"I have learned the secret of being content in any and every situation… I can do all this through Him who gives me strength."

— Philippians 4:12–13

Paul's words hit like truth grenades in the soul of anyone who's ever tried to hold it all together with broken hands. Content in any and every situation? That kind of peace sounds impossible when you're knee-deep in trauma, battling addiction, or haunted by war. Paul wasn't speaking theory, he was speaking from prison. He had nothing left in the tank… except Christ, and that was enough.

I used to think strength meant never needing help. That contentment came from having control, but I learned the hard way through combat, addiction, betrayal, and failure, that all those things can be stripped away in an instant, and when they are, what's left? For a long time, I chased peace like it was a moving target. I searched for it in drugs, women, adrenaline, money, motorcycles, and brotherhood. Every time I got my hands on what I thought would fix me, it crumbled like dust. I didn't need more stuff. I needed a Savior. I didn't need another high, I needed healing, and that's what I found in Christ. Not because I figured it all out. Not because I cleaned up my act. Because Jesus met me in my mess. He didn't wait for me to get strong, He came to me in my weakness, and I began to see, like Paul, that the secret of contentment isn't changing your circumstances, it's knowing your identity in Christ and trusting His presence in the middle of the storm.

"You don't need peace from the world when the Prince of Peace lives inside you."

When everything else is gone when the uniform's off, the medals collect dust, the marriage is broken, the bottles are empty, the mask slips, He is still there, and He is enough.

Enough when you're unemployed. Enough when you're doubting. Enough when you're back in the pit. Enough when you've buried your best friend. Christ is the bread when your soul is starving. Christ is the light in your blackout. Christ is the voice saying, "You're still Mine." He's not just enough to save you, He's enough to sustain you.

Chapter 55

Colossians

Christ, the Supreme and Sufficient One

Chapter 55. Colossians

Christ, the Supreme and Sufficient One

"He is before all things, and in Him all things hold together."

— Colossians 1:17

When the World Feels Like It's Falling Apart

We all have moments maybe whole seasons when it feels like everything's unraveling. For veterans, first responders, those in recovery, or anyone carrying hidden wounds, life can feel like a war zone long after the actual battle ends.

Paul wrote Colossians to a church being pulled in different directions some chasing rituals, others chasing secret wisdom, all at risk of forgetting the one thing that truly matters: Christ is enough.

"When the world falls apart, the One who holds all things together doesn't."

The Fullness of God in Human Form

"For in Christ all the fullness of the Deity lives in bodily form, and in Christ you have been brought to fullness."

— Colossians 2:9–10

This isn't abstract theology it's rescue theology. Paul isn't writing to impress philosophers, he's writing to anchor the broken. If Jesus is fully God and He is, then there's no part of your life that's too dark, too messy, or too shattered for Him to enter and heal.

I used to believe I needed to "get it together" before God would come near, but Colossians flips that lie on its head. God didn't wait for me to fix myself. He came in the flesh, entered the fight, bore the wounds, and brought His fullness to my emptiness. The same Christ who conquered death now dwells in you, not when you're strong, but even when you're on the floor, choking on the weight of memories and shame.

The Invisible Made Visible

"The Son is the image of the invisible God, the firstborn over all creation."

— Colossians 1:15

Want to know what God is like? Look at Jesus. He is not some distant force, cosmic judge, or angry commander. He's the God who touches lepers, washes feet, weeps at graves, and walks toward the cross.

As a veteran, I was trained to look for strength, toughness, endurance, and control, but Christ's strength looked different: it bled, it broke, it served, and that shattered my idea of what true power really is. Christ is not just some lofty theological concept, He's the image of God we can see, feel, and know. He doesn't just show us God, He shows us ourselves made whole in Him.

Buried and Raised with Him

"Having been buried with Him in baptism… you were also raised with Him through your faith in the working of God, who raised Him from the dead." — Colossians 2:12

I've been buried under shame, under addiction, and under the weight of grief, and maybe you have too, but the message of Colossians is this: you don't stay buried. In Christ, we rise.

Recovery is resurrection in motion. Healing is the soil where new life grows. We don't just die to our past we're raised to a new future, and that resurrection isn't a someday thing. It's a right now reality. You have been raised. Present tense.

Set Your Heart on Things Above

"Since, then, you have been raised with Christ, set your hearts on things above, where Christ is." — Colossians 3:1

This isn't about escaping the world. It's about living in it differently. When trauma wants to pull you into despair… when addiction whispers in the night… when the noise of politics, pressure, and pain surrounds you, Colossians gives you a compass. Look up. Christ is on the throne, and you are seated with Him. He's not ignoring your struggle. He's interceding in it. You may feel like you're drowning in the moment, but He sees the whole map. Set your heart where your hope lives, in Him.

Not visit. Dwell.

There's a difference between reading Scripture and being rooted in it. Between singing worship and letting it rewrite your identity. Colossians calls us not to consume religion, but to let Christ take up residence in every room of our hearts. When He moves in, fear moves out.

Whatever You Do, Do It for the Lord

"And whatever you do, whether in word or deed, do it all in the name of the Lord Jesus…" — Colossians 3:17

This is where faith meets daily life. Washing dishes? Worship. Welding steel? Worship. Working a recovery step? Worship. Changing diapers? Worship. You don't need a pulpit to preach. You don't need a platform to serve. In Christ, every task becomes sacred, and every scar becomes a testimony.

Chapter 56
1st & 2nd Thessalonians

Christ, Our Coming King and Steady Hope

Chapter 56. 1st and 2nd Thessalonians

Christ our Coming King and Steady Hope

"The Lord himself will come down from heaven… And so we will be with the Lord forever. Therefore encourage one another with these words."

— 1 Thessalonians 4:16–18

"But the Lord is faithful, and He will strengthen you and protect you from the evil one."

— 2 Thessalonians 3:3

Letters Written to the Weary and Waiting

Paul's letters to the Thessalonians were written to a young church under pressure. They were being persecuted, misunderstood, and tempted to give up hope. Some were grieving. Others were confused about Christ's return. Many were just tired, tired of waiting, tired of suffering, and tired of fighting.

Sound familiar?

Whether you're a veteran struggling with PTS, a first responder carrying the weight of trauma, someone in recovery from addiction, or just a believer holding on by a thread, these letters are for you. Because hope isn't just something you wait for. Hope is something you hold on to.

Anchored by the Return of Christ

The heartbeat of both letters is this: Jesus is coming again.

"For the Lord himself will come down from heaven, with a loud command, with the voice of the archangel…" — 1 Thess. 4:16

That's not just theology. That's fuel for the fire. That's strength for the next step. That's comfort for the soldier mourning a fallen brother. That's a lighthouse for the addict fighting the waves of relapse. Christ is not done. He's coming. Not to crush, but to complete. Not to shame, but to save, and we don't wait in despair. We wait in confidence.

Grief That Holds Hope

"We do not grieve like the rest of mankind, who have no hope."

— 1 Thessalonians 4:13

Grief is real. Veterans know that. We've lost friends, some in battle, some after it. Some took their own lives. Others drifted away in silence. We grieve. We cry, and We remember. But in Christ, we don't grieve like the world does. Because Jesus rose, so will they. Because Jesus conquered death, death doesn't have the final word. Our grief is laced with the promise of reunion, resurrection, and restoration.

"Tears may flow, but they don't mean it's over. They mean we still love. And love, in Christ, is eternal."

Steady in the Waiting

"May He strengthen your hearts so that you will be blameless and holy in the presence of our God…"

— 1 Thess. 3:13

Waiting is hard. Whether you're waiting for healing, for sobriety to feel stable, for relationships to mend, or just for the weight to lift, it's easy to start wondering if God forgot. But

Paul reminds us: God doesn't waste the waiting. He strengthens, he purifies, he holds us firm when everything else feels shaky.

2 Thessalonians tells us that the "man of lawlessness" may rise (2 Thess. 2), and the world may get darker, but the people of God stand not in fear, but in faith. Because our hope isn't wishful thinking, it's anchored in a risen King who keeps His promises.

Faithful in the Fight

"But the Lord is faithful, and He will strengthen you and protect you from the evil one." — 2 Thess. 3:3

Paul doesn't sugarcoat the struggle. Spiritual warfare is real. Addiction doesn't back down easy. Trauma doesn't heal overnight, but the Lord is faithful.

You are not fighting alone.

Whether it's the battle in your mind, the brokenness in your family, or the weight of your past, God is your shield. He will strengthen you. He will protect you and He will finish the work He started in you.

Don't Grow Weary in Doing Good

"And as for you, brothers and sisters, never tire of doing what is good." — 2 Thessalonians 3:13

Sometimes the hardest thing is to keep going. To keep loving people who are hard to love. To keep walking in recovery when relapse whispers. To keep showing up when you're running on fumes, but Paul says: don't give up. Your kindness matters. Your sobriety matters. Your prayers matter and your quiet acts of love matter. Because God sees, and your faithfulness is never wasted.

Chapter 57 1st & 2nd Timothy

Christ, Our Trainer in Grace and Truth

Chapter 57. 1st and 2nd Timothy

Christ, our Commander and Shephard of the Wounded

"Fight the good fight of the faith. Take hold of the eternal life to which you were called…"

— 1 Timothy 6:12

"I have fought the good fight, I have finished the race, I have kept the faith."

— 2 Timothy 4:7

Letters from a Battle Tested Mentor

Paul's letters to Timothy are different. They're not written to a church, but to a person, a young pastor, a spiritual son, and a fellow soldier in the faith. These letters are personal, gritty, and urgent. Paul is writing from experience, from wounds, and from nearing the end of his own mission. He's writing to pass on what matters most.

If you've ever sat across from an older vet, hearing stories and getting hard-earned wisdom handed down like a folded flag, that's the feel of these letters. They're battlefield briefings, not just for pastors, but for anyone trying to stay faithful in the fog of war.

A Young Leader, A Heavy Load

Timothy was young, timid, and carrying a weight bigger than himself. Paul doesn't flatter him, he fortifies him.

"For the Spirit God gave us does not make us timid, but gives us power, love and self-discipline."

— 2 Timothy 1:7

That verse hit me hard the first time I read it during recovery. I didn't feel powerful. I didn't feel loving. I didn't feel disciplined. I felt like a kid in a war I didn't sign up for, spiritually shellshocked and scared I wouldn't make it through.

But Paul is saying: you don't have to feel strong. You've already been given what you need. The same is true for us. Veterans. First responders. Recovering addicts. Survivors of trauma. When we feel disqualified, God says: "I qualified you. I called you. And I'm not letting go."

The Good Fight of Faith

Paul tells Timothy to "fight the good fight." Not with fists or weapons, but with faith.

"Endure hardship with us like a good soldier of Christ Jesus."

— 2 Timothy 2:3

We know what it's like to fight the wrong fights trying to prove ourselves, silence the guilt, and outrun the past, but Paul says there's one fight worth everything: the fight of faith. It's not easy. Faith is hard when your marriage is strained. Faith is hard when the bottle calls louder than your Bible. Faith is hard when your mind keeps reliving a day you can't undo, faith is hard, but Paul reminds us: you're not fighting alone. Christ is not only with you, He's leading you. He is our Commander, not just barking orders but entering our wounds, walking through the fire, and promising to finish what He started.

Watch Your Life and Doctrine Closely

Paul tells Timothy something that speaks right into the heart of recovery:

"Watch your life and doctrine closely. Persevere in them, because if you do, you will save both yourself and your hearers."

— 1 Timothy 4:16

It's not about legalism it's about awareness, about walking sober-minded. About being vigilant not just with our behavior, but with our beliefs. Because trauma lies, addiction lies, and guilt lies. They tell us we're unworthy, unloved, and unable to change. Sound doctrine says otherwise. It says: "You are a child of God. You are sealed by the Spirit. You are not defined by your past, and you are covered in Christ's righteousness."

Watching your life? That's not pressure it's protection. Paul isn't telling Timothy to perform. He's telling him to pay attention. Stay awake. Stay close to grace. Stay in community and stay on mission.

Discipleship Is Legacy

Paul wasn't just mentoring Timothy he was modeling the kind of man Timothy could become.

"What you heard from me, keep as the pattern of sound teaching, with faith and love in Christ Jesus."

— 2 Timothy 1:13

This is legacy work. Not just recovery, but redemption. Not just surviving, but shepherding others. That's what this book is about. Not just helping you find peace, but helping you pass that peace on. Helping you become the kind of man or woman who says to the next broken soul: I've been where you are, and Christ met me there.

Veterans, your scars don't disqualify you from ministry, they qualify you to minister to the scarred.

A God Who Guards What We Entrust

Paul says something stunning near the beginning of his second letter:

"I know whom I have believed, and am convinced that He is able to guard what I have entrusted to Him until that day."

— 2 Timothy 1:12

Have you entrusted your pain to Jesus? Have you entrusted your trauma, your losses, your regrets? Paul says: He's able to guard it, to hold it, and to keep it safe. Even when everything else is falling apart. That's peace, not because everything is fixed, but because everything is in His hands.

Finishing Well

Near the end, Paul gives what feels like his final salute:

"I have fought the good fight, I have finished the race, I have kept the faith."

— 2 Timothy 4:7

That's what we all want, isn't it? Not perfection. Not applause. Just faithfulness. To be able to say: I didn't give up. I kept showing up. I let grace carry me farther than I thought I could go, and then he writes:

"Now there is in store for me the crown of righteousness…"

— 2 Timothy 4:8

That's what's waiting for you, brother. Sister. Soldier. And Survivor. Not shame, not rejection, a crown, a reward and a welcome home.

Chapter 58

Titus

Christ, Our Trainer in Grace and Truth

Chapter 58. Titus

Christ, our Trainer in Grace and Truth

"For the grace of God has appeared that offers salvation to all people. It teaches us to say 'No' to ungodliness and worldly passions, and to live self-controlled, upright and godly lives in this present age…"

— Titus 2:11–12

A Letter to a Trusted Brother

Titus isn't just a theological pamphlet, it's a call to action. It's written to a man on the frontlines of a rough mission field: the island of Crete. A place where lies ran deep, culture was hostile, and truth was scarce. Paul doesn't coddle Titus. He commissions him, and in just three short chapters, Paul hands over battle plans for how to build a church, confront false teaching, live out integrity, and most of all, how to let the grace of Christ transform people from the inside out.

Grace That Trains Like a Drill Sergeant

This is the heartbeat of Titus:

"The grace of God… teaches us." — Titus 2:11–12

Grace is not soft. Grace is not weak. Grace is not just a get out of jail free card. Grace trains us. The Greek word for "teaches" here (παιδεύω, paideuō) is the same word used for disciplined instruction, like training a soldier. Grace is not just forgiveness, it's formation. It gets in your face, it builds spiritual muscle, and it teaches us to say no to the cravings that once ruled us and yes to a new way of life in Christ.

For veterans and first responders, this makes sense. We know the value of discipline, of repetition, of transformation under pressure. The same is true in faith. Grace doesn't just accept you as you are, it reshapes you into who you were meant to be.

Truth in a World of Lies

Paul doesn't sugarcoat Crete. He quotes a local saying:

"Cretans are always liars, evil brutes, lazy gluttons."

— Titus 1:12

In other words, the culture was corrupt, violent, and apathetic. Sound familiar? Today's world might not look so different. False gospels abound. Quick fix spiritual junk food is everywhere, and in recovery, it's easy to get lost in voices that say "do what feels good" or "you'll never change" But Paul says:

"You must teach what is appropriate to sound doctrine."

— Titus 2:1

Truth matters. Doctrine matters. Not to puff up our pride, but to anchor our lives and to tell us who God really is. Who we really are, and how we really heal. For trauma survivors and those in recovery, this is lifesaving. The truth of the Gospel cuts through the lies of shame, despair, addiction, and failure. It exposes false teaching and plants our feet on solid ground.

Leadership from the Inside Out

Paul spends a significant chunk of Titus outlining the kind of leaders the church needs: not just charismatic, not just competent, but Christlike.

"An elder must be blameless... not overbearing, not quick-tempered, not given to drunkenness, not violent, not pursuing dishonest gain."

— Titus 1:6–7

Why does this matter for a book about peace and recovery? Because many of us have been wounded by bad leadership. We've seen hypocrisy. We've been let down by authority figures, maybe even in church. Titus reminds us: God doesn't just call leaders, He transforms them.

You may not think of yourself as a leader, but if you've walked through trauma and met Christ in the valley, you have something to offer others. Leadership starts with integrity, humility, and grace fueled courage.

Our Hope: Christ, the Appearing Glory

Paul anchors everything in a future hope:

"While we wait for the blessed hope—the appearing of the glory of our great God and Savior, Jesus Christ."

— Titus 2:13

We live in the "in between," between Christ's first coming and His return. Between the trauma we've survived and the healing still in progress. Between what we've lost and what we'll one day gain, but we are not left alone. Christ has appeared, bringing grace, and Christ will appear again bringing glory. That's peace, that's strength, that's the fuel to keep going.

You Are Not Who You Were

The clearest expression of the Gospel in Titus might be this:

"At one time we too were foolish, disobedient, deceived and enslaved by all kinds of passions and pleasures... But when the kindness and love of God our Savior appeared, He saved us..."

— Titus 3:3–5

Did you hear that? We were enslaved… but now we are free. We were foolish… but now we walk in wisdom. We were condemned… but now we are justified. None of it is because of us.

"…not because of righteous things we had done, but because of His mercy."

— Titus 3:5

For those of us who have stumbled, fallen, relapsed, or ran,we need that sentence engraved on our hearts. You are not saved because of your resume. You are saved because of Christ's mercy.

Chapter 59

Philemon

Christ, Our Liberating Advocate

Chapter 59. Philemon

Christ, our Liberating Advocate

"Perhaps the reason he was separated from you for a little while was that you might have him back forever no longer as a slave, but better than a slave, as a dear brother."

— Philemon 1:15–16

A Personal Letter with Eternal Weight

Philemon is the shortest of Paul's letters, just 25 verses. But don't let its size fool you. This letter is a holy powder keg. It's packed with truth, tension, grace, and a revolutionary view of what the Gospel can do in the broken systems of the world.

It's a letter about a runaway slave named Onesimus. It's a letter to a man named Philemon, who had every legal right to punish him. It's also a letter from Paul, who calls both men to a higher standard, not of law, but of love. This is a letter that challenges every chain, literal or invisible, that still holds us, and it points us to Christ, the One who stood in the gap for all of us.

From Fugitive to Family

We don't know all the details, but it seems Onesimus fled from Philemon, possibly stealing from him in the process. Somehow, this runaway slave ends up in Rome, where he meets Paul and more importantly, meets Jesus, and then everything changes.

Paul writes:

"Formerly he was useless to you, but now he has become useful both to you and to me."

— Philemon 1:11

The name Onesimus actually means "useful." What a beautiful twist. The Gospel redeems not just his soul, but his name, his story, and his identity. Veterans, addicts, prisoners, and the broken know what it's like to be labeled "useless." But Christ doesn't throw people away. He redeems them. He rewrites their names.

Paul the Intercessor: A Picture of Christ

What Paul does next is staggering. He sends Onesimus back to Philemon but not as a slave, not as property, and not as a problem to be dealt with. He sends him back as a brother.

"If you consider me a partner, welcome him as you would welcome me."

— Philemon 1:17

Paul steps into the place of Onesimus. He says, "If he owes you anything, charge it to me." That's not just friendship, that's substitution. This is a powerful picture of what Jesus does for us. We were rebels, runaways, and spiritually bankrupt, but Christ stood in the gap and said, "Put it on My tab. I'll take the punishment. Welcome them like you'd welcome Me." That's not just forgiveness. That's reconciliation.

The Gospel Breaks Chains

Make no mistake Philemon lives in a world where slavery is normal, but Paul doesn't endorse that system. Instead, he gently detonates it from the inside.

"No longer as a slave, but better than a slave, as a dear brother."

— Philemon 1:16

Paul doesn't organize a rebellion he speaks a new reality into existence. He calls Philemon to treat Onesimus not according to the law of Rome, but according to the law of Christ. This hits home for anyone who's ever been bound by addiction, shame, legalism, past mistakes, or generational wounds. Christ doesn't just break chains He replaces them with bonds of brotherhood.

A Call to Costly Grace

Paul doesn't command Philemon, he appeals to him:

"Though in Christ I could be bold and order you to do what you ought to do, yet I prefer to appeal to you on the basis of love."

— Philemon 1:8–9

Grace always invites. It doesn't force, but that doesn't make it cheap. This kind of love costs something. For Philemon, it might mean loss of pride or property. For Onesimus, it means risk, walking back into the unknown, trusting that the Gospel really does change hearts.

For you and me, grace might mean forgiveness we don't want to give… or trusting we'll be received when we return home with our heads low, but in every case, it means Jesus goes before us. Advocating, interceding and reconciling.

Veterans, Prisoners, and Runaways

This letter matters deeply to those of us who feel like Onesimus, like we've run too far, messed up too bad, and broken too many promises. It also speaks to those who are like Philemon, those who've been wronged and now face the hard road of forgiveness, and for the ones like Paul? It reminds us that we are called to be bridges. Between people and God. Between the offended and the offender. Between what was and what could be. That's the power of the Gospel: it doesn't erase the past. It redeems it.

Chapter 60 Paul From Persecutor to Preacher

Chapter 60. Paul from Persecutor to Preacher

A Journey of Redemption

"But the Lord said to Ananias, 'Go! This man is my chosen instrument to proclaim my name to the Gentiles and their kings and to the people of Israel.'"

— Acts 9:15

The Road to Damascus and the Road to Redemption

Paul's story is the ultimate comeback. He wasn't just a skeptic he was an enemy of the Church. Known then as Saul of Tarsus, he breathed threats, arrested believers, and stood in silent approval as Stephen was murdered. He was a man fueled by law, driven by zeal, and blind to the grace of God, and then Christ broke through the darkness.

On the road to Damascus, Saul was knocked off his high horse, literally and spiritually. A flash of divine light. A voice from heaven. "Saul, Saul, why do you persecute me?" (Acts 9:4). With those words, the Lord of glory laid claim to the heart of a man who thought he was doing everything right.

That moment of confrontation became the turning point. Saul the persecutor became Paul the preacher. The man who once tried to silence the Gospel became the one who would carry it farther than anyone else in history.

From Enemy to Ambassador

Veterans and first responders often carry heavy burdens memories of things done, seen, or left undone. Guilt can weigh like armor. Shame can whisper that you'll never be more than your worst day, but Paul's life tells a different story.

He called himself "the worst of sinners" (1 Tim. 1:15). He was honest about his past, but he didn't stay there. He didn't let his former identity define his future. Christ had written a new name over his life: Chosen Instrument.

Maybe your story started in rebellion, addiction, violence, or despair. That's not where it has to end. If God can use a man like Saul, He can use anyone, including you.

A Mission Fueled by Grace

Paul didn't just get saved, he got sent. Once blinded, he now saw the Gospel with crystal clarity. Once chained to law, he now breathed the air of grace. His letters, Romans, Corinthians, Galatians, Ephesians, Philippians, Colossians, Thessalonians, Timothy, Titus, Philemon are love letters from a battlefield theologian, written to churches under pressure.

Everywhere Paul went, he preached Christ crucified, our righteousness, our hope, and our peace. He was beaten, stoned, shipwrecked, imprisoned, and betrayed. He wept, he wrestled, and he bled, but he never backed down. Not because he was strong, but because grace made him bold.

"But by the grace of God I am what I am, and his grace to me was not without effect."

— 1 Corinthians 15:10

Christ Is Enough

In 2 Corinthians 12, Paul pleads with God to remove a "thorn in the flesh." We don't know what it was physical pain, spiritual torment, depression, PTS like symptoms, but we do know God's answer:

"My grace is sufficient for you, for my power is made perfect in weakness."

— 2 Corinthians 12:9

That verse has comforted countless warriors in their darkest moments. When the trauma doesn't lift. When the addiction still lingers. When the guilt keeps revisiting. Christ whispers the same truth: "I am enough."

Paul's Final Days: Faithful to the End

Paul's last known words are preserved in his second letter to Timothy. He writes from a Roman prison, knowing death is near. Yet listen to his tone, not one of regret, but of resolve:

"I have fought the good fight, I have finished the race, I have kept the faith."

— 2 Timothy 4:7

That's not just a preacher's farewell. That's a soldier's signoff. Paul's legacy wasn't comfort or applause. It was faithfulness through pain, persecution, and prison.

That's the call to us today: Not to be perfect, but to be faithful. Not to be fearless, but to be full of the Spirit. To run our race, scars and all, toward the finish line where Christ waits to crown us not with medals, but with mercy.

Chapter 61
Hebrews

Christ, Our Great High Priest and Unshakable Anchor

Chapter 61. Hebrews

Christ, our Great High Priest and Unshakable Anchor

"We have this hope as an anchor for the soul, firm and secure. It enters the inner sanctuary behind the curtain, where our forerunner, Jesus, has entered on our behalf…"

— Hebrews 6:19–20

Hope for the Soul That's Drifting

The book of Hebrews was written to a people weary from the fight persecuted, misunderstood, tempted to give up and go back. Many were considering walking away from Christ because the pressure felt too great. They had left the old ways, but the new life felt fragile and uncertain. If you've ever thought, "I don't know if I can keep doing this" Hebrews is for you.

Hebrews is a lifeline for the disillusioned believer, the hurting veteran, the recovering addict, and the trauma survivor. It reminds us that Jesus is not only the sacrifice for our sin, but also the priest who brings us near, the brother who walks beside us, and the King who reigns over us.

Christ Is Better

The key word in Hebrews is better, better than angels, better than Moses, better than the old covenant, and better than the blood of bulls and goats.

"The Son is the radiance of God's glory and the exact representation of his being… After he had provided purification for sins, he sat down at the right hand of the Majesty in heaven."

— Hebrews 1:3

Veterans and first responders know how tempting it is to look back to the "old way" of numbing pain, stuffing trauma, hiding behind a hardened heart, but Hebrews tells us, Don't go back. What you have now in Christ is better. The old systems were shadows. Jesus is the substance.

A High Priest Who Understands

One of the most comforting truths in Hebrews is that Christ is not distant or detached. He gets it.

"For we do not have a high priest who is unable to empathize with our weaknesses, but we have one who has been tempted in every way, just as we are—yet he did not sin."

— Hebrews 4:15

When you're struggling with guilt, when the memories haunt you at night, when shame whispers, "You're unworthy" you don't have to hide. Christ isn't shocked. He isn't repulsed. He's been there in the trenches of human pain, and now, He stands before the Father interceding for you—not as a cold judge, but as your advocate and brother.

An Anchor for the Soul

Storms come. Some of us have been through physical storms combat zones, broken homes, and addiction battles. Others face the mental and spiritual kind, depression, anxiety, PTS, and spiritual fatigue.

Hebrews 6 offers us this promise:

"We have this hope as an anchor for the soul, firm and secure..."

— Hebrews 6:19

Jesus is the Anchor. Not our performance. Not our toughness. Not even our faith because sometimes that faith wavers, but He never does. His grip holds when ours fails. This anchor has gone "behind the curtain," into the very presence of God. Christ went ahead of us through the cross, and through the grave and now holds our place in glory.

A Better Sacrifice

The old covenant demanded sacrifice after sacrifice, day after day, year after year. But Hebrews 10 gives us a radical truth:

"But when this priest had offered for all time one sacrifice for sins, he sat down at the right hand of God."

— Hebrews 10:12

Why did He sit down? Because the work was finished.

Veteran, you don't have to atone for your past. You don't have to bleed out your guilt in silence. Christ already did. Once. For all.

Hold Fast

Over and over, Hebrews urges us: Don't give up. Don't drift. Don't turn back. Not because you have to muscle through, but because Christ is holding you.

"Let us hold unswervingly to the hope we profess, for he who promised is faithful."

— Hebrews 10:23

We all have days where faith feels fragile, but Jesus never asks you to be perfect only to come to Him. He is faithful even when we're not.

Faith That Endures

Hebrews 11 is known as the Hall of Faith. It lists people who didn't have easy lives they had trials, loss, exile, war, and martyrdom. Yet they endured by faith.

"They were longing for a better country a heavenly one."

— Hebrews 11:16

You might not feel like a hero of faith, but faith isn't about perfection. It's about perseverance. Even when you stumble. Even when you limp. Hebrews 12 says to fix your eyes on Jesus, "the pioneer and perfecter of our faith." He started this journey in you. He'll finish it too.

Chapter 62
James

Christ, Our Wisdom in the Fire

Chapter 62. James

Christ, our Wisdom in the Fire

"Consider it pure joy, my brothers and sisters, whenever you face trials of many kinds, because you know that the testing of your faith produces perseverance."

— James 1:2–3

Straight Talk for Wounded Warriors

James doesn't sugarcoat anything. There's no theological fine print, no vague philosophy. Just truth raw, practical, and piercing. He speaks like a drill sergeant who's seen combat. His words cut through excuses and aim straight for the heart.

That's why this book hits home for veterans, first responders, and anyone in recovery. James doesn't coddle. He challenges, but his challenge is rooted in hope: real change is possible because Christ lives in us.

This book isn't for the religiously polished, it's for those who've been in the trenches. For those who know how hard life can hit. For those who are tired of just talking about faith and are ready to walk it out.

Faith That Works in the Real World

James doesn't just ask what you believe he asks what you do with it.

"Faith by itself, if it is not accompanied by action, is dead."

— James 2:17

This isn't about earning salvation it's about living like it matters. Real faith gets its hands dirty. It shows up in how we treat people, how we speak, and how we endure pain. For those of us who once lived with masks on performing, pretending, or hiding, James offers a better way: authenticity. Not perfection, but a life of real, gritty, Spirit filled obedience.

Trials That Shape Us, Not Break Us

James opens with a wild statement: "Consider it joy when you face trials." At first glance, it feels impossible. Joy in trauma? Joy in suffering? James isn't saying to like the pain he's saying to trust the process. Trials reveal what's underneath. They surface our doubts, fears, and cracks, but they also expose how deeply Christ is forming something eternal in us. For the veteran with combat flashbacks, the addict in recovery, or the mom trying to hold her family together, James says: God's not done with you. The fire is forging something stronger.

"Blessed is the one who perseveres under trial…"

— James 1:12

Watch Your Mouth It Matters

James devotes an entire section to the power of words:

"The tongue is a small part of the body, but it makes great boasts… it is a restless evil, full of deadly poison."

— James 3:5, 8

Veterans know the damage a single order or phrase can do. In trauma and addiction recovery, we learn that words shape our reality what we tell ourselves, what we speak over others, an what we believe about God. James invites us to let the Spirit tame our tongues, to speak life, not death. To speak healing, not harm and to speak blessing, not bitterness.

Wisdom from Above, Not Below

There's a difference between street smarts and spiritual wisdom. James describes two kinds: Earthly wisdom: fueled by pride, selfishness, and bitterness. Heavenly wisdom: pure, peace loving, gentle, submissive, full of mercy and good fruit.

"Peacemakers who sow in peace reap a harvest of righteousness."

— James 3:18

In a world driven by outrage, violence, and self-justification, Christ calls us to be peacemakers. Not doormats, but warriors of gentleness, men and women who carry the authority of heaven by the fruit of the Spirit.

Humble Yourself You're Not God

James reminds us that pride is poison:

"God opposes the proud but shows favor to the humble."

— James 4:6

Veterans and leaders often struggle with this. We were trained to take charge, push through, and "never quit." But James says humility is the only path to peace. When we stop trying to control everything… when we lay down our need to prove something… when we surrender even our healing process to Christ, that's where the breakthrough happens.

Don't Just Plan Pray

"Now listen, you who say, 'Today or tomorrow we will go to this or that city…' Why, you do not even know what will happen tomorrow."

— James 4:13–14

We make plans. We map out recovery steps, deployments, life goals. But James reminds us: God is still in charge. This doesn't mean we shouldn't plan, it means we should anchor those plans in prayer. Always saying, "If it is the Lord's will…" Because peace doesn't come from controlling the future, it comes from trusting the One who holds it.

God Sees. God Hears. God Will Restore.

James closes with assurance for the suffering:

"Be patient, then, brothers and sisters, until the Lord's coming… The Lord is full of compassion and mercy."

— James 5:7, 11

For those waiting for justice… waiting for healing… waiting for restoration… James says: Don't give up. God hasn't forgotten you.

He sees the hidden tears. He remembers the trauma others ignored., and He promises mercy that never runs dry.

Chapter 63

1st & 2nd

Peter

Christ, Our Cornerstone and Coming King

Chapter 63. 1st and 2nd Peter

Christ our Cornerstone and Coming King

"Cast all your anxiety on Him because He cares for you."

— 1 Peter 5:7

"But in keeping with His promise we are looking forward to a new heaven and a new earth, where righteousness dwells."

— 2 Peter 3:13

A Letter to the Scattered and Suffering

Peter, once the brash fisherman who denied Christ in fear, writes these letters as a man transformed by grace and refined by fire. He writes to people scattered by persecution—displaced, disoriented, and discouraged. In many ways, they were spiritual exiles, much like how many veterans, trauma survivors, and those in recovery often feel: strangers in a world that doesn't understand their pain.

These letters are not casual encouragements they are battlefield communiqués. Peter speaks as a fellow soldier of the cross, offering hope to the hurting and strength to the suffering.

Christ Our Living Hope

Peter begins his first letter with a powerful reminder:

"Praise be to the God and Father of our Lord Jesus Christ! In His great mercy He has given us new birth into a living hope through the resurrection of Jesus Christ from the dead."

— 1 Peter 1:3

For those who have lost hope whether in a hospital room, a courtroom, a foxhole, or in the grip of addiction Peter offers more than wishful thinking. He offers living hope, anchored in the resurrection. This isn't about positive thinking. This is about a living Savior who defeated death and walks with us through our darkest days.

Suffering with Purpose

"Though now for a little while you may have had to suffer grief in all kinds of trials… these have come so that the proven genuineness of your faith… may result in praise, glory and honor when Jesus Christ is revealed."

— 1 Peter 1:6–7

Peter doesn't tell us to pretend the pain isn't real. He tells us that suffering has a purpose. Like a soldier refining his skills under fire, or a recovering addict learning to lean on God in the valley, our trials are not wasted. God uses them to forge a faith that is genuine, tested, and resilient.

A Call to Holy Living

Peter's call to holiness isn't about shame or legalism, it's a call to identity. He reminds us:

"But you are a chosen people, a royal priesthood, a holy nation, God's special possession…"

— 1 Peter 2:9

For those who have felt discarded by the world veterans returning to civilian life, felons trying to rebuild, those whose pasts are marked by addiction and trauma, Peter speaks a better word: You belong. You are not forgotten. You are chosen.

Christ Our Cornerstone

Peter calls Jesus the "living Stone" rejected by humans but chosen by God.

"See, I lay a stone in Zion, a chosen and precious cornerstone, and the one who trusts in Him will never be put to shame."

— 1 Peter 2:6

When your life feels like it's collapsing, when trust is shattered, when foundations are cracked, Christ is the Cornerstone. He doesn't shift under pressure. He doesn't fail when everything else does. Build your life on Him.

Warfare, Discipline, and Self Control

"Be alert and of sober mind. Your enemy the devil prowls around like a roaring lion looking for someone to devour."

— 1 Peter 5:8

Peter speaks our language combat, vigilance, awareness. He reminds us we have an enemy, and that enemy isn't just external, it's spiritual. The trauma, the addictions, the despair, they're part of a deeper war, but we're not left defenseless. We're called to stand firm in faith, with the promise that God Himself will restore, confirm, strengthen, and establish us (1 Peter 5:10).

Cast Your Burdens

For the veteran who wakes up in sweat-soaked sheets, for the first responder haunted by what they've seen, for the mom in recovery who fears she'll lose her kids again, Peter says:

"Cast all your anxiety on Him because He cares for you."

— 1 Peter 5:7

This isn't a spiritual cliché, it's a lifeline. The Commander of heaven cares. The weight you carry is not too heavy for Him. Cast it. Don't carry it alone.

2 Peter: Holding the Line

Years later, Peter writes again. He's older now. Wiser. A man who knows he is nearing his final deployment, his execution under Nero is coming, but before he goes, he reminds the Church to hold the line.

Christ's Power Is Enough

"His divine power has given us everything we need for a godly life..."

— 2 Peter 1:3

We often feel underequipped. Not spiritual enough. Not strong enough, but Peter says Christ's power has already given us everything we need. Everything. Not someday, but now. His Spirit, His promises, His Word, and His presence. They're enough to face the shame, the cravings, the trauma, and the doubts.

Growth Is the Mission

Peter lays out a path of growth:

"Make every effort to add to your faith goodness... knowledge... self control... perseverance... godliness... mutual affection... and love."

— 2 Peter 1:5–7

This is not a ladder to climb but a journey to walk. A healing path for the veteran who's rebuilding. A roadmap for the addict learning to live again. A battle plan for the believer choosing peace over panic. We don't grow by trying harder. We grow by abiding deeper.

False Teachers and Forgotten Promises

Peter warns of deception of voices that pull us away from truth. And he calls out the question so many silently ask:

"Where is this 'coming' He promised?"— 2 Peter 3:4

In other words: Why hasn't God fixed everything yet? Peter answers with both truth and tenderness. God is not slow. He's patient. He's giving people time to repent, but the Day is coming. Justice will come. The King will return.

A New Heaven and New Earth

"In keeping with His promise we are looking forward to a new heaven and a new earth, where righteousness dwells."

— 2 Peter 3:13

The war will end. The pain won't last. The story isn't finished. There is a coming world renewed, healed, and righteous. Until then, we stand firm and grow in grace.

Chapter 64

1st, 2nd, & 3rd John

Christ, Our Love, Light, and Assurance

Chapter 64. 1st, 2nd, and 3rd John

Christ, our Love, Light, and Assurance

"See what great love the Father has lavished on us, that we should be called children of God! And that is what we are!"

— 1 John 3:1

"God is love. Whoever lives in love lives in God, and God in them."

— 1 John 4:16

"I have no greater joy than to hear that my children are walking in the truth."

— 3 John 1:4

The Apostle of Intimacy and Identity

By the time John writes these letters, he's an old man possibly the last living apostle. He has outlived many of his fellow disciples, endured exile, persecution, and heartbreak, but his message hasn't grown bitter. It's grown bolder in love.

John doesn't write as a theologian debating from a distance. He writes like a father who's watched generations drift, suffer, and question their worth. His voice is full of tenderness but also holy fire. He is calling God's people back to the basics: truth, love, and assurance.

For the veteran, the weary soul, the recovering addict, and the traumatized believer John's words are a call home.

God Is Light Walk in the Light

"God is light; in Him there is no darkness at all."

— 1 John 1:5

John doesn't pretend darkness isn't real. But he reminds us where our hope is found. Not in hiding, not in pretending, but in stepping into the light.

"If we confess our sins, He is faithful and just and will forgive us… and purify us from all unrighteousness."

— 1 John 1:9

This isn't a shaming light it's a healing light. For the combat veteran haunted by moral injury, for the former addict riddled with guilt, for the survivor afraid of their story this verse is freedom. You can bring it all into the open, and when you do, you'll find not judgment, but faithful forgiveness.

Children of God Beloved and Named

"See what great love the Father has lavished on us, that we should be called children of God!"

— 1 John 3:1

John's entire framework of faith is relational. Not performance driven. Not shame based. Not feared. He speaks to the soul that's been crushed by war, addiction, or rejection and says: You are not what you've done. You are not what's been done to you. You are His.

Your truest identity is not "broken veteran" or "recovering addict" it is child of God, and when that sinks in, peace begins to replace panic. Purpose begins to rise from pain.

Christ's Love Perfect, Fearless, Unfailing

"There is no fear in love. But perfect love drives out fear."

— 1 John 4:18

Fear tells us we're not enough. That we'll be abandoned. That our worst day defines us, but John's message dismantles fear at its root: Love casts it out. Christ's love is not insecure or temperamental. It's fearless. Total. Steady.

For those who were raised in fear, who served in violence, or who survived trauma, this kind of love may feel foreign. But John assures us: this is the love God has for you. It is not earned. It is not fragile. It is lavished.

Test the Spirits Guard the Truth

"Do not believe every spirit, but test the spirits to see whether they are from God..."

— 1 John 4:1

John doesn't just talk about love he talks about truth. He knows that in a broken world, there are many false voices: shame, legalism, lies from the enemy, religious pride. Some sound spiritual but lead to chains.

Veterans know this, too. Not all commands are righteous. Not all voices are trustworthy. John calls us to discern. To weigh what we hear against the truth of Jesus Christ crucified, risen, and full of grace and truth.

2 John: Truth and Love in Balance

In his second letter, John writes to a woman and her children likely a house church. It's a short letter, but it packs a punch:

"I ask that we love one another. And this is love: that we walk in obedience to His commands."

— 2 John 1:5–6

Here, John connects love to obedience not as a burden, but as the natural outflow of knowing Christ. We don't follow Jesus out of fear, we follow Him because we've encountered His love. And love compels obedience, not rebellion.

John also warns against those who "do not continue in the teaching of Christ" (v. 9). Love is not blind it is discerning. It guards the truth because the truth sets us free.

3 John: Walking in the Truth

In his final letter, John commends Gaius for being faithful:

"I have no greater joy than to hear that my children are walking in the truth."

— 3 John 1:4

It's a reminder that the Christian life isn't just about head knowledge it's about walking. Living. Moving in truth.

For the veteran rebuilding after addiction, for the father trying to restore trust with his children, for the woman in recovery learning to stand again, walking in the truth may start slow, but every step matters.

Chapter 65
Jude

Christ, Our Keeper and Contender for the Faith

Chapter 65. Jude

Christ, our Keeper and Contender for the Faith

"To Him who is able to keep you from stumbling and to present you before His glorious presence without fault and with great joy."

— Jude 1:24

"Dear friends, although I was very eager to write to you about the salvation we share, I felt compelled to write and urge you to contend for the faith that was once for all entrusted to God's holy people."

— Jude 1:3

A Battle for Truth and Hope

Jude, the brother of James and likely the half brother of Jesus, writes one of the shortest books in the New Testament but it hits like a hammer. It's a call to spiritual vigilance, a reminder that faith must be fought for, and a declaration that Christ will keep His people to the end.

For those who've served in combat, this letter sounds familiar. It's not written from a place of comfort or theological abstraction. It's written with urgency. With grit. With eyes wide open to the spiritual battle raging all around.

Contend for the Faith

"I felt compelled to write and urge you to contend for the faith..."

— Jude 1:3

Jude wanted to write about salvation, but the Holy Spirit redirected him. Why? Because there were enemies within the camp false teachers distorting grace, turning it into a license for sin. This wasn't just misinformation it was a betrayal of Christ. To "contend" for the faith means to wrestle, to fight, to refuse to surrender truth for comfort. Veterans know this instinct: hold the line. That's Jude's heartbeat. Don't let your faith be eroded. Stand your ground in grace and truth.

Combat Faith, Not Just Comfort Faith

We live in a world that wants a faith that's easy, that conforms to our feelings and culture, but Jude reminds us that real faith is forged in battle. It's not about being loud or angry, it's about being anchored.

"They are ungodly people, who pervert the grace of our God into a license for immorality and deny Jesus Christ…"

— Jude 1:4

The enemy doesn't always wear a uniform. Sometimes he whispers lies in the church pew. Sometimes he tempts you to trade conviction for convenience. Jude isn't calling us to paranoia but to discernment.

Christ Our Keeper

One of the most powerful truths in Jude is this: You are kept.

"To those who have been called, who are loved in God the Father and kept for Jesus Christ."

— Jude 1:1

The same Jesus who bled for you now guards you. The One who defeated death is the One who holds your faith. You're not keeping yourself saved by willpower, Christ is keeping you. He is your Keeper in the battle.

That's good news for those who feel too tired to hold on. For the soldier fighting PTS, for the woman recovering from years of trauma, for the addict trying not to relapse again Jude says: He will keep you.

Remember the Past, Stand in the Present

Jude draws heavily from Israel's history reminding his readers how even angels fell, how Sodom was judged, how Cain rebelled. He's not just pointing fingers; he's warning: Don't forget the cost of compromise.

"But you, dear friends, by building yourselves up in your most holy faith and praying in the Holy Spirit, keep yourselves in God's love…"

— Jude 1:20–21

This is your spiritual briefing: build yourself up, pray with passion, stay rooted in God's love, and wait with expectancy for His mercy.

Mercy for the Broken

Even as Jude warns of judgment, he reminds us to show mercy:

"Be merciful to those who doubt; save others by snatching them from the fire…"

— Jude 1:22–23

Not everyone in danger wants to be saved, but some are just barely hanging on. Maybe that was you once. Maybe it's you now. Veterans know what it's like to pull someone from danger. Those in recovery know what it's like to sit beside someone on the brink. Jude is reminding us: the mission includes rescue. Don't just defend the faith extend the hand of Christ.

The Final Benediction: You're Not Alone

Jude ends with one of the most powerful doxologies in all of Scripture:

"To Him who is able to keep you from stumbling…"

— Jude 1:24

That's the anchor. You are not keeping yourself by discipline or perfection you're being kept by a Savior who doesn't sleep on post. A Savior who sees your scars and still calls you His. A Commander who won't leave you behind.

Chapter 66

Revelation

Christ, the Victorious King and Faithful Witness

Chapter 65. Revelation

Christ, the Victorious King and Faithful Witness

"Look, He is coming with the clouds, and every eye will see Him…"

— Revelation 1:7

"They will wage war against the Lamb, but the Lamb will triumph over them because He is Lord of lords and King of kings."

— Revelation 17:14

A Battle Already Won

Revelation is not just the end of the Bible it's the unveiling of ultimate reality. It's not primarily a secret code or timeline it's a promise. A vision. A battle report from heaven itself, and at the center of it all is Jesus Christ, not as a meek teacher or broken Savior, but as the Risen Warrior King.

For veterans, first responders, and anyone who has fought through darkness, Revelation isn't intimidating, it's familiar. It's a battlefield. It's a call to endure, but more than anything, it is the declaration that the war is already won.

Hope in the Fire

Revelation was written to persecuted believers exiles, outcasts, prisoners, and martyrs. John, writing from the island of Patmos, had seen war, death, and trauma firsthand. He had buried brothers and lived through heartbreak. He wrote Revelation not as an escape, but as a lifeline of hope.

"I, John… was on the island of Patmos because of the word of God and the testimony of Jesus."

— Revelation 1:9

John was a prisoner of war for the Gospel, and from that exile, he sees heaven open. If you've ever felt banished by your trauma, cut off by your addiction, silenced by your shame John's vision is for you.

The One Who Walks Among Us

The first vision is not dragons or beasts, it's Christ Himself, walking among the lampstands (the churches). Eyes like fire. Voice like rushing waters. Face shining like the sun.

"Do not be afraid. I am the First and the Last. I am the Living One; I was dead, and now look, I am alive forever and ever!"

— Revelation 1:17–18

This is not a distant God. This is a Commander who walks the front lines. He is among us even when the night is long and the silence is deafening.

To the One Who Overcomes

Revelation 2–3 contains letters to seven churches some thriving, some barely hanging on. Each one ends with a promise:

"To the one who overcomes…"

These aren't promises for the perfect. They're for the battle worn. For the faithful. For the ones who don't quit.

- "I will give the right to eat from the tree of life."

• "I will give you hidden manna… a white stone with a new name."

• "You will not be hurt at all by the second death."

• "I will give you authority over the nations."

• "I will never blot out your name."

• "You will be a pillar in the temple of my God."

• "You will sit with Me on My throne."

These are warrior promises. And they are yours in Christ.

Worship in the Throne Room

At the center of Revelation is a throne not of chaos, but of glory. John is pulled into a heavenly war room filled not with fear, but with worship.

"Holy, holy, holy is the Lord God Almighty, who was, and is, and is to come."

— Revelation 4:8

Then he sees a Lamb, slain but standing. All of heaven erupts in praise.

"Worthy is the Lamb who was slain, to receive power and wealth and wisdom and strength…"

— Revelation 5:12

For the one who has been broken by war, by loss, by addiction this is your center. Not the battlefield. Not the trauma. Not the shame. The throne of grace, where a Lamb rules in majesty.

The Dragon and the Blood of the Lamb

Yes, Revelation speaks of beasts and battles, dragons and deception. But the real fight isn't for power it's for allegiance and the victory doesn't come by swords or strength. It comes by the blood of the Lamb and the word of your testimony.

"They triumphed over him by the blood of the Lamb and by the word of their testimony..."

— Revelation 12:11

That means your story matters. Your scars are part of the war journal. Every time you choose hope, every time you speak the name of Jesus instead of drowning in silence, you overcome.

The New Heaven, The New Earth

Revelation ends not in rubble, but in resurrection. Not in terror, but in tenderness.

"He will wipe every tear from their eyes. There will be no more death or mourning or crying or pain..."

— Revelation 21:4

This is not fantasy. This is our future. Veterans haunted by what they've seen this is your healing. Survivors of abuse this is your justice. Addicts and the ashamed this is your redemption. Jesus doesn't just bring peace He is Peace.

Chapter 67

Christ in Every Chapter

Chapter 67. Christ in Every Chapter

Peace in Every Page

"And beginning with Moses and all the Prophets, He explained to them what was said in all the Scriptures concerning Himself."

— Luke 24:27

We have walked a long road together. From the breath of God over the chaos in Genesis to the final trumpet blast of Revelation. From the wilderness of Exodus to the exile of Daniel, from the poetry of David to the passion of Christ, from broken prophets to bold apostles—we have found Christ in every book.

This was never about just gathering information or completing a devotional checklist. This was about rediscovery. About tracing the thread of a Redeemer who never left us, even when we ran. It was about seeing how every shadow, every symbol, every miracle, every promise, every name, points to Jesus.

And not just the Jesus we read about on paper but the Living Christ who walks with us, weeps with us, and brings peace to the war zones of our souls.

The One Story That Holds All Stories

There are 66 books in the Bible. Written over 1,500 years. Penned by kings and shepherds, prophets and fishermen, prisoners and poets, and yet there is one Author behind it all. One storyline that holds the whole thing together:

God's relentless pursuit of us through Jesus Christ.

Every law in Leviticus was a shadow of the One who would fulfill it.

Every king in Chronicles was a flicker of the true King to come.

Every sacrifice in the temple whispered of the Lamb of God.

Every Psalm cried out for the Shepherd.

Every judgment echoed the cross.

Every promise hinted at the Resurrection.

This story is not just ancient, it is active.

It's not just true, it is life changing.

From Battlefield to Sanctuary

For veterans like us, the Bible doesn't feel foreign, it feels familiar.

- We know the cost of sacrifice.
- We know what it means to lose brothers.
- We understand wilderness, trauma, guilt, and grief.
- We've tasted shame.
- We've numbed ourselves to forget.
- We've stood alone in crowded rooms, unsure if we belonged.

And yet… Christ was in all of it.

He was with us in the desert.

He wept with us in the barracks.

He heard us in the silence.

He stood between us and the bullet more times than we know.

This book has shown us that peace doesn't come from running, escaping, numbing, or pretending.

Peace comes when we meet Christ in the Scriptures.

Realize He's been chasing us from Genesis 1.

Hope for the Wounded

The enemy wants you to believe you're too far gone. That because of what you've done, or what's been done to you, God's not interested. That you're damaged goods.

But Scripture says otherwise.

- In Genesis, we found the God who covers our shame.
- In Judges, we met a Deliverer who works through weakness.
- In Psalms, we cried out and were heard.
- In Isaiah, we were promised a Suffering Servant.
- In the Gospels, we saw that Servant crucified and risen.
- In Acts, we joined a mission bigger than ourselves.
- In Romans, we were justified.
- In Corinthians, we were made part of the body.
- In Ephesians, we were seated in heavenly places.
- In Revelation, we stood victorious with the Lamb.

There is no book in Scripture that doesn't welcome the wounded.

There is no scar that excludes you from the story.

Peace That Surpasses Understanding

The peace we've been chasing through this journey isn't cheap.

It was bought with blood.

It was earned by Christ, not us.

It is kept by His Spirit, not our strength.

And it's ours, now and forever.

"Therefore, since we have been justified through faith, we have peace with God through our Lord Jesus Christ."

— Romans 5:1

That's not a maybe. That's not a wishful thought. That's a verdict.

No matter what the war inside you looks like today, Christ is your Peace.

No matter how far you've wandered, Christ is your Way Home.

No matter how broken you feel, Christ is your Healer.

No matter how many times you've failed, Christ still calls you His own.

Devotional Reflection: The Story Isn't Over

This book may be nearing its final pages, but your story is not finished.

You are not too late. You are not too lost. You are not too broken.

You are part of a greater story, one written in the ink of grace, folded in the pages of Scripture, and sealed by the Spirit of Christ Himself.

Keep going. Keep reading. Keep walking with Him. You haven't just found Christ in every book, You've found Him in your story too, and that means there is always hope, always peace, always purpose, and always a next step.

Let this not be the end, but the beginning. Let Your Word remain alive in me. Let Your Spirit continue to heal me. Let Your peace define me from this day forward.

Amen.

Bibliography

Throughout this book you read some great quotes from some of the most recognized theologians, pastors, religious leaders and authors of our time. They are listed below in the order they appear in my story.

J.I. Packer

Dallas Willard

John Eldredge

A.W. Tozer

Brad Powell

Bishop T.D. Jakes

Robert Wolgemuth

C.S. Lewis

Dietrich Bonhoeffer

Anne Lamott

Scripture Translations Used:

New International Version (NIV)

English Standard Version (ESV)

New Living Translation (NLT)

***Additional Resources Referenced*:**

Veterans Crisis Line – www.veteranscrisisline.net

Mighty Oaks Warrior Programs – www.mightyoaksprograms.org

Celebrate Recovery – www.celebraterecovery.com

National Alliance on Mental Illness (NAMI) – www.nami.org

U.S. Department of Veterans Affairs PTSD Resources – www.ptsd.va.gov

Made in the USA
Monee, IL
12 November 2026

f5f26be4-2455-40c9-a011-60eae6e0e782R02